FOOD INDUSTRY BRIEFING SERIES:
SHELF LIFE

For Lai Ching and Keziah

FOOD INDUSTRY BRIEFING SERIES: SHELF LIFE

Dominic Man

Principal Lecturer in Food Sciences at the School of Applied Science,
South Bank University, London

Blackwell
Science

Editorial Offices:
Osney Mead, Oxford OX2 0EL
25 John Street, London WC1N 2BS
23 Ainslie Place, Edinburgh EH3 6AJ
350 Main Street, Malden
MA 02148 5018, USA
54 University Street, Carlton
Victoria 3053, Australia
10, rue Casimir Delavigne
75006 Paris, France

Other Editorial Offices:

Blackwell Wissenschafts-Verlag GmbH
Kurfürstendamm 57
10707 Berlin, Germany

Blackwell Science KK
MG Kodenmacho Building
7–10 Kodenmacho Nihombashi
Chuo-ku, Tokyo 104, Japan

Iowa State University Press
A Blackwell Science Company
2121 S. State Avenue
Ames, Iowa 50014-8300, USA

First published 2002

Set in 10/13 Franklin Gothic Book
by DP Photosetting, Aylesbury, Bucks
Printed and bound in Great Britain by
TJ International Ltd, Padstow, Cornwall

DISTRIBUTORS

Marston Book Services Ltd
PO Box 269
Abingdon
Oxon OX14 4YN
(*Orders:* Tel: 01235 465500
Fax: 01235 465555)

USA and Canada
Iowa State University Press
A Blackwell Science Company
2121 S. State Avenue
Ames, Iowa 50014-8300
(*Orders:* Tel: 800-862-6657
Fax: 515-292-3348
Web: www.isupress.com
email:
orders@isupress.com)

Australia
Blackwell Science Pty Ltd
54 University Street
Carlton, Victoria 3053
(*Orders:* Tel: 03 9347 0300
Fax: 03 9347 5001)

A catalogue record for this title is available from the British Library

ISBN 0-632-05674-6

Library of Congress
Cataloging-in-Publication Data
Man, Dominic.
Shelf life / Dominic Man.
p. cm.--(Food industry briefing series)
Includes bibliographical references and index.
ISBN 0-632-05674-6
1. Food--Storage. 2. Food--shelf-life dating.
I.Title.II.Series.

TP373.3 .M36 2001
664'.028--dc21 2001052650

For further information on
Blackwell Science, visit our website:
www.blackwell-science.com

Contents

Series Editor's Foreword

All food businesses today operate in complex technical and commercial environments and, if they are to survive, each food business must possess the skills and knowledge required to remain ahead of competitors. Many businesses, and particularly the larger businesses, employ staff who are specialists in the diverse fields that support the food industry's enterprise: from food science and technology, microbiology and engineering to marketing management, logistics and operations management. Yet, though a business may be well served by highly qualified and capable staff, expert in their given field, the needs of the business may not always be fully met. When new skills are required an option is to recruit new staff with the appropriate expertise, but even for the larger business this may not be the preferred option. For the smaller business it is often out of the question. An alternative, and one frequently taken today, is to give staff the development opportunity to learn about the subjects and develop the skills required. In this way food businesses can efficiently and effectively acquire the expertise they need to develop and grow, and to operate within the increasingly demanding frameworks of the established law.

Of the problems faced by staff who desire to learn about new subjects for job development – or who just want to dip into new fields to broaden their knowledge – the choice of study material can be a significant barrier. With so much information available today choosing the right material to study and finding material that can be quickly and easily assimilated can become a complex and off-putting task. In recognition of these issues, the *Food Industry Briefing Series* was devised to assist the food industry in staff development and to provide a useful resource for food industry staff. It is intended primarily for the use of executives, managers and supervisors, but should also find application in academia. Each volume will cover a given topic related to the activities of the food industry and each will aim to provide the essence of the subject matter for ready assimilation either for use in its own right or to create a foundation upon which to layer the concepts contained in more academically demanding texts. The *Food Industry Briefing Series* is intended to make it easy for the reader to become conversant with, and develop a practical understanding of a particular subject, such that the information and ideas gained can be immediately and confidently applied in any setting, from the factory floor to the boardroom.

Shelf Life, by Dominic Man, is, of course, about the shelf life of foods. The book is written by an acknowledged expert in the field and will be an invaluable source of information and ideas for those working on shelf life, or for anyone wanting to develop their knowledge of the subject area. As a text which will appeal to both the food industry and academia, it is very well written, concise and a pleasure to read. Though relatively short (which was intentional) it is packed with knowledge communicated in a way that makes it immediately, and practically, useful. The organoleptic qualities of food are a cornerstone of competitive advantage in the marketing of food, and are significantly influenced by issues concerned with shelf life. This book is likely, therefore, to become a standard reference source for an industry which is highly sensitised to the need always to satisfy consumers' tastes and to meet their expectations of the way food quality is maintained and food keeps.

Ralph Early
Series Editor, *Food Industry Briefing Series*
Harper Adams University College
October 2001

Preface

The importance of shelf life to all concerned in the manufacture, processing, distribution, sale and consumption of food in today's society cannot be questioned. It has therefore been a great honour to be involved in the compilation of this book.

In keeping with the overall aim of this *Food Industry Briefing Series*, the intention is to provide a concise and quick reference book for the busy food industry professionals, particularly those working in small and medium-sized enterprises, whose responsibilities include the determination and management of shelf lives of their products. The book is also aimed at senior undergraduate students reading food science, food technology or a related discipline. Recent graduates in their first jobs practising as food scientists or technologists may also find this book a useful reminder of what they have learned.

The book was put together during a difficult period for the author, both professionally and personally. I would like to thank my family for their love and encouragement that have sustained me through this time. I have, too, to thank Nigel Balmforth, the publisher, for not giving up on me and for his wise counsel and exceptional patience. My heartfelt gratitude goes to Rosa Pawsey and Michael Hoffmann, who, at very short notice, kindly read and made helpful comments on the draft manuscript. Reviewing the huge amount of high-quality publications on which this book is largely based has been an exciting, humbling and rewarding endeavour. I sincerely hope you will find reading this book a worthwhile experience.

C.M.D. Man
South Bank University
London

Section 1

Introduction to Shelf Life of Foods – Frequently Asked Questions

1.1 What is shelf life?

Shelf life is a frequently used term that can be understood and interpreted differently. A consumer is generally concerned with the length of time a food product can be kept in the home before it can no longer be used. A retailer is particularly interested in the length of time a product can stay on its shelf in order to maximise sales potential. Here, shelf life is defined as the period of time under defined conditions of storage, after manufacture or packing, for which a food product will remain safe and be fit for use. In other words, during this period, it should retain its desired sensory, chemical, physical, functional or microbiological characteristics and, where appropriate, comply with any label declaration of nutritional information when stored according to the recommended conditions (IFST, 1993). It is obvious, therefore, that shelf life is a very important and multifaceted requirement of all

Table 1.1 Examples of food and drink product recalls extracted from various national newspapers in the UK between 1992 and 2001

Product	Published reason for the recall
Red and white wines	Did not meet quality and safety specifications; alleged to be contaminated with methyl isothiocyanate
Dairy fruit desserts/yoghurt	Glass contamination
Vacuum packed cooked chestnuts	Faulty packaging
Crispy chicken	Contained small pieces of red plastic
Cauliflower with lamb babymeal	Contained a residue of a cleansing agent in excess of the UK advisory level

(Contd)

Table 1.1 *(Contd)*

Product	Published reason for the recall
Traditional roasted ham	Did not meet Health and Safety standards
Snack (meat) pies	Metal contamination
Widget beers	Loose piece of plastic widget might be present
Ready-to-pour gravy	Contamination could potentially pose a risk to health
Summer fruit meringue roulade	Praline and chocolate meringue roulade, containing hazelnuts, mistakenly packed in the wrong cartons
Potato crisps	Might contain pieces of glass
Mince pies	Contaminated with small pieces of rubber
Vegetable lasagnes	Vegetable bake, containing cashew nuts, wrongly packaged
White sliced bread	Contained small slivers of metal wire
Carbonated soft drinks	Contained traces of benzene due to contaminated carbon dioxide
Chicken yakitori	Processing fault led to a small quantity being undercooked
Canned chopped tomatoes	Contained higher than acceptable levels of tin
Milk chocolate raisins	Contained chocolate peanuts
Canned long cut spaghetti in tomato sauce	Some contained higher than acceptable levels of tin
Sparkling wine brut	Contained tiny particles of glass
Organic vegetable soup	Might have received inadequate heat treatment
Toffee crisp ice cream	Might contain traces of nuts not mentioned on packaging
Coarse farmhouse pâte	A small number had been inadequately heat treated
Chilled snackpot-style convenience products	Plastic packaging ignited during microwave reheating
Canned tomato soup	Might contain levels of tin that exceed statutory limit
No-added-sugar orange, lemon and pineapple squash	Might be unsuitable for consumption due to an excess of one ingredient
Canned baked beans	Defective side seam which may have meant they were not effectively sealed
Ready-to-eat whole roast chicken	Due to a processing fault, a limited number may not have been fully cooked

manufactured and processed food products. Every food product has, and should be recognised as having, a microbiological shelf life, a chemical shelf life and an organoleptic shelf life, because all foods deteriorate, albeit at different rates. Ultimately, the shelf life of a food product is intended to reflect the overall effect of these different aspects, ideally under a set of specified storage conditions. Because of this, the study of shelf life of food can often only rightfully be dealt with by the employment of multidisciplinary resources.

1.2 Why are food safety and shelf life related?

The safety of food is both a fundamental and legal requirement. It follows that all food products offered for sale must be safe although they do not necessarily have to be of the highest quality. In the UK, the *Food Safety Act* (HMSO, 1990; Anon., 1996) prohibits the sale of food that:

- Has been rendered injurious to health.
- Is unfit.
- Is so contaminated it would be unreasonable to expect it to be eaten.
- Is not of the nature or substance or quality demanded.
- Is falsely or misleadingly labelled.

Table 1.1 gives a list of past food product recalls in the UK; this should give some insight into the kinds of problems which responsible food businesses consider to be hazardous. In effect, a food product, the safety of which has been called into question, has no useful shelf life. Food safety and product shelf life are therefore inextricably linked. Without exception, the question 'Is this product safe to eat?' must precede every shelf life determination. As every product or product concept has to be taste tested at some stage, it is only right and proper for ethical reasons to resolve this question at the earliest opportunity. Furthermore, the controlling factors for safety and spoilage, particularly those that are related to microbial growth, are often identical; the separate consideration of safety and shelf life, although convenient in practice, is artificial. Today, the most effective way to ensure the safety of food is, of course, to use the internationally recognized Hazard Analysis Critical Control Point (HACCP) system.

1.3 Who should be interested in shelf life of foods?

Since shelf life is such an important requirement, it should be of interest to everyone involved in the food chain. There is a growing realization that a high standard of food safety and quality can only be achieved by adopting a comprehensive and integrated approach, covering the whole of the food chain 'from farm to table'. As will be seen later (see Section 1.6), there are many factors that can influence the shelf life of food. The use of a cleaner ingredient in an ambient cake filling (for example, roasted chopped almonds as opposed to chopped almonds), which has a lower microbial load, could mean a difference between an acceptable rather than an unacceptable shelf life for the cake as a whole. Suppliers of raw materials and ingredients, and food manufacturers and producers, can often overcome potential shelf life problems by working closely together at the earliest opportunity. At the other end of the food chain, consumers, too, have a significant part to play. For instance, by minimizing the exposure of foods to high temperatures, particularly during summer months, and by observing carefully any recommended storage and usage instructions, consumers are ensuring that the intended shelf lives of their foods will not be reduced.

1.4 Who is responsible for determining shelf life?

Basically, the responsibility for determining shelf life lies with the manufacturer or the packer. While ideas for new products and for improvements to existing products can originate from within a food business and from external sources such as a current or prospective customer, shelf life evaluation and testing are very much integral parts of every product development programme. Therefore, it is in keeping with the established principles of good manufacturing practice (GMP) that a food manufacturer should possess its own in-house shelf life testing and evaluation capability (Blanchfield, 1998). Today, almost without exception, major retailers do independently evaluate the shelf lives of food products, particularly their own-label ones. This, however, should neither negate nor reduce the responsibility of a food manufacturer or processor whose duty it is to assign correct shelf lives to their products based on experimental work which has been carried out.

1.5 Is it illegal to give a wrong shelf life to a food product?

In the UK, one of the principal provisions of the *Food Labelling Regulations* (HMSO, 1996) that applies to all food which is ready for delivery to the ultimate consumer or to a catering establishment, subject to certain exceptions, is that it should be marked or labelled with the appropriate minimum durability indication (FSA, 2000). Thus one of the following must be given:

- In the case of a food which, from the microbiological point of view, is highly perishable *and* in consequence likely after a short period to constitute an immediate danger to health, a 'use by' date.
- In the case of a food other than one specified above, an indication of minimum durability, a 'best before' date.

Additionally, the 'best before' date and the 'use by' date must be followed by any special storage conditions which need to be observed, such as 'keep refrigerated at 0°C to +5°C' or 'keep in a cool, dry place'. Generally, storage conditions are important because in the EU's main food labelling Directive 79/112/EEC (EEC, 1979), which the *Food Labelling Regulations* (HMSO, 1996) implement, the date of minimum durability is defined as the date until which the foodstuff retains its specific properties when properly stored. Guidance is available on what foods should carry a 'use by' date. Foods that require a 'use by' date are likely to fall into the following categories (Crawford, 1998):

- Dairy products, e.g. dairy based desserts.
- Cooked products, e.g. ready-to-eat meat dishes, sandwiches.
- Smoked or cured ready-to-eat meat or fish, e.g. hams, smoked salmon fillets.
- Prepared ready-to-eat foods, e.g. vegetable salads such as coleslaw.
- Uncooked or partly cooked pastry and dough products, e.g. pizzas, sausage rolls.
- Uncooked products, e.g. uncooked products comprising or containing either meat, poultry or fish.
- Vacuum or modified atmosphere packs, e.g. raw ready-to-cook turkey breast packed in modified atmosphere.

In practice, the range of products which are given a 'use by' date differs from country to country (Mröhs, 2000). In the end it is the manufacturer's and processor's responsibility to decide to which category their product belongs and whether a 'use by' or 'best before' date is the appropriate indication. In general, the date must be given as a day, month and year, in that order. For the 'best before' date category, the following forms of durability indication are allowed (Mröhs, 2000):

- Foods that will not keep for more than 3 months – 'best before' followed by the day and the month.
- Foods that will keep for more than 3 months but not more than 18 months – 'best before end' followed by the month and the year.
- Foods that will keep for more than 18 months – 'best before end' followed by the month and the year or the year only.

Since food deteriorates continually rather than suddenly, the 'best before' date does not automatically mean the food is not fit for consumption or loses all its acceptability immediately after that date.

Once a date mark (either 'use by' or 'best before') is set and declared, it becomes a contract between the food company and its customers to the effect that, provided the food is stored according to the recommended conditions, it should last at least as long as its stated shelf life. In order to be confident of its statement, the company must have done the necessary work to determine the correct shelf life; the marking of a product with either a 'use by' or 'best before' date does imply that this is so. It follows that giving a wrong date mark (i.e. shelf life) to a food product will make the company liable to enforcement actions. Indeed, in the UK, in the event of a complaint resulting from a product that has spoiled before the end of its declared shelf life, the company may well be asked by an enforcement agency to justify the validity of the shelf life claimed. The corollary is that objective evidence in the form of comprehensive records of shelf life study could legitimately be used by the company as part of a due diligence defence as provided for in the *Food Safety Act* (HMSO, 1990).

Retailers have different responsibilities depending on whether or not the products are branded or own-labelled ones. Besides carefully observing stock rotation procedures, some retailers use an additional voluntary 'display until' date to ensure that high-risk chilled foods, in particular, are not on display beyond their 'use by' date. A supermarket in the UK was fined some years ago for selling out-of-date (i.e. 'use by' date) products, so the penalty for any errors could be significant.

The EU's food labelling Directive 79/112/EEC (EEC, 1979) has so far been supplemented by three further Directives, 94/54/EC (EC, 1994), 96/21/EC (EC, 1996) and 99/10/EC (EC, 1999). Relevant to the issue of shelf life, the first (Directive 94/54/EC; EC, 1994) requires that the words 'packaged in a protective atmosphere' should appear on the labels of foodstuffs whose shelf life has been extended by means of packaging gases (see Section 1.13). This requirement applies to all products packaged using modified atmosphere packaging (MAP). A useful introduction to the requirements of the UK food labelling regulations, including the minimum durability indication requirement, hosted by the University of Reading, is available at: http://www.fst.rdg.ac.uk/foodlaw/label/index.htm.

1.6 How long a shelf life should my product have?

How long is a piece of string? There is really no straightforward answer.

All foods spoil and they do so differently and at different rates; even for the few exceptions such as some wines and cheeses, the acceptability of which improves on storage (i.e. maturation/ripening), their quality invariably deteriorates once their optimal acceptability has been reached. Despite the enormous range and variety of food products available nowadays, much knowledge about food deterioration has been accumulated. Although one must guard against generalization, most food spoilage can be explained by one or more of the following mechanisms (IFST, 1993):

- Moisture and/or water vapour transfer leading to gain or loss.
- Physical transfer of substances other than moisture and/or water vapour, e.g. oxygen, odours or flavours.
- Light-induced changes, i.e. changes caused and/or initiated by exposure to daylight as well as artificial light.
- Chemical and/or biochemical changes.
- Microbiological changes.
- Other mechanisms or changes that cause the food to deteriorate through one or more of the above-mentioned mechanisms, e.g. damage to the pack caused by insect infestation.

Furthermore, temperature, the single most important environmental factor, influences all these mechanisms, so the effects of temperature must be evaluated in all shelf life studies.

Knowing the spoilage mechanism of a food product, therefore, is the first step in the process of determining its shelf life. Essentially, how a food spoils and hence how long its shelf life is are going to be influenced by a number of factors. These shelf-life influencing factors are the properties of the final product and of the environment in which it is to be manufactured, stored, distributed and used. These factors can be divided into the following groups:

1. Intrinsic factors (see Section 2.2.1):
   - Raw materials.
   - Product composition and formulation.
   - Product structure.
   - Product make-up.
   - Water activity value (a_w).
   - pH value and total acidity.
   - Availability of oxygen and redox potential (E_h).
2. Extrinsic factors (see Section 2.2.2):

- Processing.
- Hygiene.
- Packaging materials and system.
- Storage, distribution and retail display (in particular with respect to exposure to light, fluctuating temperature and humidity, and elevated or depressed temperature and humidity).

3. Other factors:
   - Consumer handling and use (see Section 2.2.4).
   - Commercial considerations (see Section 2.2.5).

Additionally, interactions between intrinsic and extrinsic factors are possible. For example, the interaction of factors such as water activity, pH, salt, nitrite and storage temperature in controlling the growth of *Clostridia* in cured meat is well known (Roberts & Gibson, 1986). Because levels of established preservatives (e.g. salt, nitrite, sugars, sorbic acid) have been reduced in many traditional products in response to consumer/market demands, to the extent that no single factor is responsible for the microbiological stability and safety of the product, it has become more and more important to understand the effects of factors acting in combination.

Further information about food deterioration mechanisms and factors affecting shelf life is given in Section 2.

1.7 What is accelerated shelf life determination?

Accelerated shelf life determination (ASLD) is used to shorten the time required to estimate a shelf life which otherwise can take an unrealistically long time to determine. As a result of globalization of food trade as well as intensification of national and international competition in the food market, the need for more rapid determination of shelf life has generally become greater. The situation is much more pressing when the shelf life of a product is expected to be long, ranging from a number of months to a few years. The effect of elevating temperature on many chemical reactions as well as adverse changes in food during storage is well known. The most common form of accelerated shelf life determination therefore relies on storing food at an elevated temperature. The assumption is that by storing food at a higher temperature, any adverse effect on its storage behaviour and hence shelf life may become apparent in a shorter time. The shelf life under normal storage conditions can then be estimated by extrapolation using the data obtained from the accelerated determination.

The following are some examples of well established accelerated storage tests:

- Incubation of canned foods for 4 or 5 days at 55°C (for the examination of thermophilic bacteria).
- Incubation of low- and medium-acid canned foods for a minimum of one week at 37°C (for the estimation of tin pick-up).
- Storage of ambient cake and pastry products at 27°C (and 75% relative humidity) (for the estimation of mould-free shelf life).
- 'Forcing' beer at 27°C (for the examination of general spoilage).
- Storage of chocolate and chocolate-coated products at 28°C (and 70% relative humidity) (for the study of bloom development).
- Accelerated storage tests such as the Schaal oven test (at 60–70°C) for the determination of edible oil stability.
- Accelerated tests at elevated pressure and temperature carried out using an instrument such as the OXIPRES™ (Mikrolab Aarhus A/S, Denmark) for the determination of oil stability in composite products such as potato crisps, margarine and biscuits, without having to extract the oil and fat from the products before analysis; the test is based on the principle of an oxygen bomb, which was utilized in the Sylvester test originally developed by J. Lyons & Co., London.

Accelerated determinations are particularly useful when the patterns of changes are practically identical during normal and accelerated storage so that shelf life under normal storage can be predicted with a high degree of certainty. For instance, it has been found that changes in quality of orange juice, made from frozen concentrate and packed in TetraBrik™, after 6 months at 20°C corresponded to the changes after 13 days at 40°C and after 5 days at 50°C (Petersen *et al.*, 1998).

Accelerated storage tests do have limitations. Essentially, they tend to be product-specific; their results have to be interpreted carefully based on detailed product knowledge and sound scientific principles. Other limitations include the following (IFST, 1993; Mizrahi, 2000):

- As temperature rises, a change of physical state may occur (e.g. melting of solid fats), which in turn can affect the rates of certain reactions.
- Although temperature is often a dominant factor and hence used as an accelerating factor, storage at a constant elevated temperature with a lower than normal relative humidity can lead to unexpected results.
- During freezing, reactants are concentrated in the unfrozen part of the food (e.g. frozen meat) resulting in a higher rate of reaction at a reduced temperature.
- A change in the way a food spoils at elevated temperatures will give false results.
- Accelerated tests are of limited use for short shelf-life chilled foods due to changes in spoilage associations at different temperatures, i.e. different storage temperatures select different spoilage microflora; besides, for short shelf-life products, the need for accelerated tests is greatly reduced.
- The Arrhenius model (see Appendix A) on which many accelerated tests are popularly based, is only appropriate for simple chemical systems and often fails for foods that are, in reality, more complex.

The most important point is that all results must be validated to confirm the relationship between changes under ASLD and those under normal storage. To be of practical use, the validated relationship should hold true at least for the product in question if not for the same product type, for instance, all tomato-based products packed in unlacquered cans (Ellis & Man, 2000).

1.8 What are the resources required for determining shelf life?

A commercially successful food product is expected to have an acceptable and reproducible shelf life. In a sense, the achievement of such a shelf life epitomises the commitment to food safety and consistent quality of the company in question. Safety and quality do not happen by chance and have to be designed into a product. The shelf life determination of foods therefore demands substantial resources, made available by management understanding and commitment. The basic resources needed are listed as follows:

1. People who possess the relevant knowledge (e.g. up-to-date knowledge in meat science and technology in a meat products company) and experience, and who can plan, carry out or supervise the evaluation, analyse the data generated and information obtained and interpret the results.
2. Adequate tools and facilities – these include (see also Section 1.14) the following:
   - Storage facility pertinent to the type of product being studied, e.g. refrigerated cabinets for chilled foods.
   - Microbiological examination facility.
   - Chemical analysis facility.
   - Sensory evaluation facility.
3. An appropriate management system that ensures every shelf life study is conducted in a systematic and timely manner, which facilitates the flow of information and communication among all those involved in it.

Although laboratory facilities for microbiological examination and chemical analysis are not absolutely essential as the required work can be undertaken by an outside laboratory, the responsibility for ensuring that the shelf life is determined accurately and reproduced consistently in production remains with the manufacturer. In the long run, it will be more cost-effective and make commercial sense to have a basic in-house shelf-life determination capability. Sometimes, specialist information and/or non-routine tests such as microbiological challenge testing or packaging migration tests are required. In this case, an external laboratory may have to be used. Without management understanding and commitment, this will not be possible.

1.9 How is the end of shelf life normally decided?

Having established the way(s) by which a food product spoils, the main task of a shelf life study is to find out as accurately as possible, under specified storage conditions, the point in time at which the product has become either unsafe or unacceptable to the target consumers. The period of time from manufacture or processing to this end-point is the maximum shelf life of the product, which has to be determined.

An end-point can be fixed with the help of:

1. Relevant food legislation, e.g. the *Dairy Products (Hygiene) Regulations* (HMSO, 1995a).
2. Guidelines given by enforcement authorities or agencies in support of their work, e.g. the UK Public Health Laboratory Service (PHLS).
3. Guides provided by independent professional bodies such as the UK Institute of Food Science and Technology (IFST), e.g. *Development and Use of Microbiological Criteria for Foods* (IFST, 1999).
4. Current industrial best practice such as that published in the primary literature.
5. Self imposed end-point, e.g. declared nutrition information such as level of an added vitamin that continues to degrade during storage.
6. Market information, e.g. results from the analysis and/or examination of a competitor's product.

Examples of some of the above are given in Table 1.2. In situations where established guidance is not available, manufacturers and processors will have to set their own end-points, using microbiological examination, chemical analysis, physical testing and, of course, properly designed and conducted sensory evaluation to define

Table 1.2 Some guidance that can be used to set shelf life end-point.

Source of guidance	Useful guidance for shelf life end-point
Food legislation	*The Dairy Products (Hygiene) Regulations 1995 (SI 1995/1086)* (HMSO, 1995a) These require a zero level for *Listeria monocytogenes* at the production stage of a dairy product (i.e. absent in 25 g for cheese other than hard cheese and absent in 1 g for other milk products) *The Tin in Food Regulations 1992 (SI 1992/496)* (HMSO, 1992) These prohibit for sale or import any food containing a level of tin exceeding 200 milligrams per kilogram. *The Coffee and Coffee Products Regulations 1978 (SI 1978/1420)* (HMSO, 1978) These require a minimum dry matter content of 95% (i.e. a maximum of 5% moisture content) for dried coffee extract (dried extract of coffee, soluble coffee, instant coffee).
Public Health Laboratory Service (PHLS)	*Guidelines for the microbiological quality of some ready-to-eat foods sampled at the point of sale* (Gilbert *et al.*, 2000)
Institute of Food Science and Technology (IFST)	*Development and Use of Microbiological Criteria for Foods* (IFST, 1999)
Published literature	It is generally accepted probiotic functional foods and drinks should contain at least 10^7 live and active bacteria per g or ml for their functional claims to be maintained over the shelf life period. (Knorr, 1998; Holzapfel *et al.*, 1998; Shortt, 1999; Birollo *et al.*, 2000)

product-specific sensory criteria (see also Section 1.14). It has to be emphasized that even established standards could change over time so that the most up-to-date ones should be used. For example, the PHLS guidelines for the microbiological quality of ready-to-eat foods was first published in 1992 and revised in 1996 before the latest guidelines became available in 2000 (Gilbert *et al.*, 2000). Likewise, the industry standard for tin in food was 250 milligrams per kilogram before the *Tin in Food Regulations* (HMSO, 1992) made 200 milligrams per kilogram a maximum level.

1.10 How do we ensure that the shelf lives established for our products are accurate and reproducible?

Shelf lives of food products are rarely established and confirmed without repeated determinations. In general, the greater the number of repeated determinations, the more accurate the assigned shelf life will be. At least four types of shelf life determinations can be distinguished, each serving slightly different purposes (IFST, 1993):

- Initial shelf life study. This is normally conducted during the concept product development stage when neither the actual production process nor the product or packaging format has been finalized. Safety of the product has either been evaluated or is evaluated alongside this study. The latter provides an indication of the probable mechanism by which the product is likely to deteriorate.
- Preliminary shelf life determination. This is the first detailed determination. It is normally carried out during the latter part of the pilot development stage or when successful plant or factory trials have been completed. Information obtained is used to assign a provisional shelf life, which will be included in the draft product, process and packaging specifications.
- Confirmatory shelf life determination. This is normally carried out towards the end of the product development process, using product samples made under factory conditions and to a set of provisional specifications. Information and data obtained are intended to confirm the provisional shelf life previously established. They will be used to finalize the provisional specifications in preparation for product launch.

  It is envisaged, however, for certain types of products such as long life ones, confirmatory determination will not be completed until after product launch. In this case, confidence in the provisional shelf life has to be based on results obtained from validated ASLD or experience derived from estimating shelf lives of established products.
- Routine shelf life determination. This is carried out in support of normal production. It provides useful information on which revision of shelf life can be based. In certain types of products

such as fresh fruits and vegetables, because of their variable nature, routine shelf life determination is an integral part of the daily packing operations. Shelf life tests are used to forewarn packers and retailers of potential quality problems, inform management regarding any shelf life adjustment, and to reveal temporal patterns in quality that can be used to trigger a change in the source of supply (Aked, 2000).

As pointed out in Section 1.6, there are many factors that can influence shelf life and so its reproducibility will be affected by many factors. A shelf life determined solely on the strength of samples that have been made by highly skilled personnel using ingredients of exceptional quality is unlikely to be reproduced exactly under factory conditions. The following factors, although not exhaustive, will need to be taken into consideration when interpreting shelf-life data generated from the different shelf life determinations (IFST, 1993):

- Trial (sample) versus bulk ingredients quantities and their range of quality.
- The age of materials used for trials and for full production.
- Variations in the weighing up of full scale formulations.
- Any scale effects as a result of scaling-up to full productions.
- Short and controlled trial runs versus fully scheduled production runs separated only by cleaning periods and/or personnel breaks.
- Batch processes versus continuous ones.
- Fluctuations of processing conditions and their full implications.
- Time factors consequent on handling full-production amounts (e.g. where product is being held longer at an elevated temperature).
- Legitimate (i.e. agreed and specified) use of surplus and/or waste materials, e.g. dough trimmings.

It is always advisable, therefore, to set the final shelf life based on data that relate to the 'worst case' manufacturing and storage scenario, and to give it a clear safety margin. This is also to recognize that because variability exists in quality of raw materials (e.g. microbial load) as well as processing conditions, there will be a distribution of shelf lives rather than an absolute shelf life that terminates abruptly and completely. In any case, the shelf life can be reviewed and if necessary, either extended or reduced in the light of further experience gained after product launch.

The secret of a reproducible shelf life that is acceptable to both the consumers and the manufacturer lies in the careful application of GMP principles. The latter, when implemented fully and effectively, will ensure the consistent manufacture of safe food products to a previously specified quality appropriate to their intended use (Blanchfield, 1998).

1.11 Can computer models help in shelf life determination?

In recent years, the widespread use of personal computers with their ever-increasing computing power has encouraged and made possible the development of computer-based models that can be used to predict the safety and shelf life of an expanding range of food products. Because of the unequivocal need to assure microbiological safety in foods, the majority of well-known computer-based models are predictive microbiological models for foodborne pathogens. Predictive food microbiology is an emerging field of study that combines elements of microbiology, mathematics and statistics to develop models, i.e. mathematical equations, that describe and predict the growth and decline of microbes under prescribed (including varying) environmental conditions (Baird-Parker & Kilsby, 1987; Fu & Labuza, 1993; McMeekin *et al.*, 1993; Whiting, 1995). Predictive microbiological models have been classified as follows (Whiting & Buchanan, 1993; McDonald & Sun, 1999):

- Primary level models. These describe changes in microbial numbers or other microbial responses (e.g. acid production, toxin synthesis) with time to a single set of conditions. Examples are the Gompertz function, *D* values of thermal inactivation and Baranyi's non-autonomous differential equation (Baranyi *et al.*, 1993).
- Secondary level models. These describe the responses of one or more parameters of a primary model to changes in one or

more of the cultural (environmental) conditions such as temperature, pH or a_w. The square root or Ratkowsky model (Ratkowsky *et al.*, 1982), response surface and Arrhenius models are examples of this class of models.

- Tertiary level models. These are computer programs (i.e. software packages) that enable users to 'interrogate' primary and secondary level models to obtain predictions.

Some of the well-known computer programs are given below.

1. Food MicroModel. This is the product of a large multicentre research project (1989–1994) initiated and funded by the UK Ministry of Agriculture, Fisheries and Food (MAFF). It is now available as single user and network versions from Food MicroModel Ltd in the UK. The package consists of a number of models, each of which is organism-specific. Most of the models are for the major foodborne pathogenic bacteria, but a number are for spoilage organisms. All models within the package have been shown to generate predictions relevant to most food groups. A prediction service based on Food MicroModel is available from the Leatherhead Food Research Association in the UK for anyone without the software to obtain predictions.
2. Pathogen Modeling Program (PMP). This is a predictive pathogen modelling program developed by the United States Department of Agriculture (USDA), which is available free on the Internet (http://www.arserrc.gov/mfs/pathogen.htm) (Fig. 1.1). The program consists of a number of models, including a few recent additions, i.e. gamma irradiation models for *Salmonella typhimurium*, *E. coli* O157:H7 and 'normal' flora in meats. The models have not been formally validated before their inclusion in the program.
3. Forecast. This is a collection of predictive models developed by the Campden and Chorleywood Food Research Association (CCFRA) in the UK, which can be used to assess the microbial spoilage rates or likely stability of foods. It is offered as a paid service by CCFRA (see Fig. 1.2).
4. *Pseudomonas* Predictor. This is a temperature function integration software developed at the Department of Agricultural Science, University of Tasmania, Australia (McMeekin & Ross,

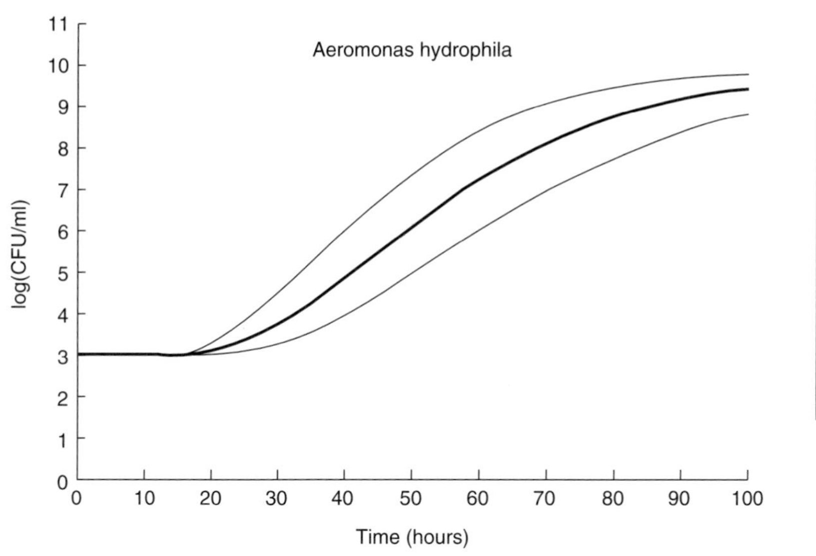

Fig. 1.1 Predictive food microbiology – predicted microbial growth curve from the USDA, ARS Pathogen Modeling Program website (http://www.arserrc.gov/mfs/pathogen.htm).

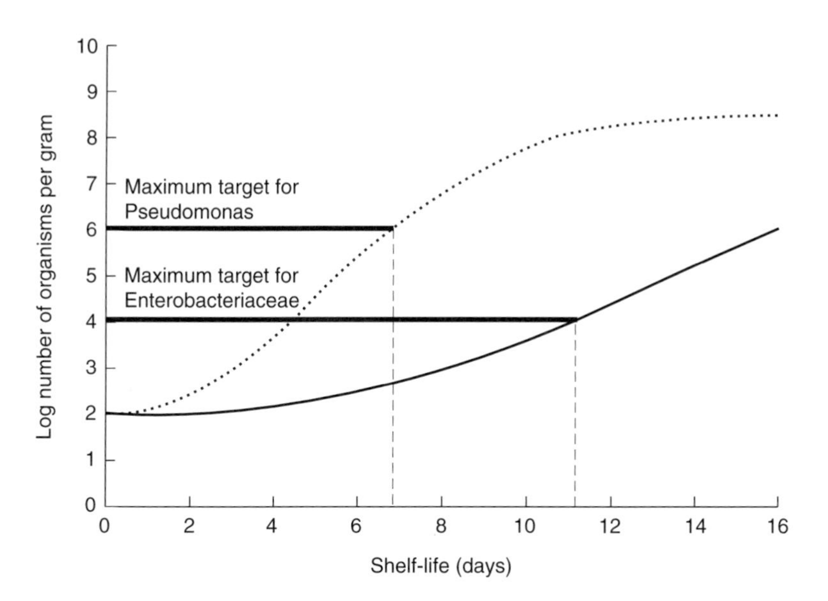

Fig. 1.2 Predicting microbiological shelf life of foods using CCFRA's Forecast service. (Reproduced with kind permission of Campden & Chorleywood Food Research Association.)

1996). It is based on work undertaken to model the effects of temperature, water activity and pH on the growth rate of psychrotrophic spoilage pseudomonads in a wide range of moist proteinaceous foods. The software has been commercialized and is marketed in Australia under the name Food Spoilage Predictor (Blackburn, 2000).

5. ERH CALC™. This is part of a computer-based 'Cake Expert System' for the baking industry originally developed by the UK Flour Milling and Baking Research Association (which is now part of the CCFRA). It allows users to run simulations on flour confectionery formulations and rapidly calculate their theoretical equilibrium relative humidities (ERHs) and hence estimate their mould-free shelf lives (MFSLs). The complete system is available from CCFRA.
6. Seafood Spoilage Predictor (SSP). This was developed at the Danish Institute for Fisheries Research (DIFRES) in Lyngby. The software contains two types of model, namely, the relative rate of spoilage (RRS) model and the specific spoilage organisms (SSO) model. It can be downloaded free of charge from the homepage of the microbiology group at DIFRES (http://www.dfu.min.dk/micro/ssp/).
7. MicroFit. This is a stand-alone software program designed to analyse microbial growth data. It was developed in the UK by the Institute of Food Research (IFR), Norwich, with funding from MAFF and four food companies. It is available as freeware and can be downloaded from the Institute's website (http://www.ifr.bbsrc.ac.uk/MicroFit/).
8. CoolVan. Available commercially since March 2000, this is a computer package for predicting food temperatures in refrigerated transport. It is the product of a UK MAFF LINK collaborative research project co-ordinated by the Food Refrigeration & Process Engineering Research Centre, University of Bristol. The program allows the user to model the effect of over a hundred variables on food temperature, including:
   - Journey details – distance, speed between deliveries, door openings, direction.
   - Weather conditions – temperature, humidity, cloud cover.
   - Food – type, amount and initial temperature.

- Van construction – insulation type and thickness.
- Type of refrigeration system.

In view of the major influence temperature has on the shelf life of chilled foods, this program is expected to make a positive contribution to the temperature control of chilled foods during distribution leading to increased confidence in their shelf lives.

The main uses of predictive microbiological models are as follows (Walker, 2000):

- New product development. Validated models can be used to assess the likely microbiological safety and stability of a product formulation. Furthermore, models will enable the following example questions that are central to shelf life determination to be answered:
  - What level of specific microorganisms will be present at various storage periods?
  - What is the effect on microbiological shelf life of reducing the salt content by 1%?
- Process design. Processing is one of the major shelf-life determining factors. With the aid of validated models for inactivation, the process can be designed to ensure that the target microorganism(s) are effectively eliminated.
- HACCP. Although HACCP has been recommended to be used only for the assurance of food safety hazards, the principles can be applied to assure product quality and shelf life. This will involve identifying the major quality hazards that influence shelf life and determining their critical control points. Validated models can be used, for example, to predict the growth potential of spoilage organisms and to estimate the tolerances that may be permitted at the various control points.
- Risk assessment. This is a rapidly developing area and models are expected to contribute towards quantitative risk assessment for the major foodborne pathogens as part of an overall effort to raise food safety standards.
- Time–temperature profiles. During storage, food products are often subject to fluctuating environmental conditions such as temperature variations. If these conditions are known, pre-

dictive models can be used to determine their cumulative effects on the microbiological shelf life of foods, especially chilled foods. Temperature function integration (TFI) has been shown to predict accurately the growth of mesophilic indicator and pathogenic microorganisms in chilled foods (McMeekin *et al.*, 1993). The technique uses the previous temperature history of the product and integrates it with the temperature-related characteristics of specific microorganisms. TFI has been applied to food storage, cooling, distribution and display (Gill, 1996).

- Training and education. Increasingly, predictive models are being used as a useful training and education tool. Used with care, they will allow food scientists and technologists to appreciate more fully how different factors such as pH, temperature and composition, can act independently as well as in combination to affect the microbiological safety and stability of food products.

Useful as they are, predictive models for food microbiology do have their limitations. In general, extrapolation cannot be made outside the ranges of factors used to produce the data in a model. Growth models will give incorrect predictions for foods that contain, for instance, natural antimicrobial substances. Also, models complement, but do not replace, the experience and skills of a food microbiologist. The ability of predictive models to indicate the microbiological shelf life of food will be limited, unless our understanding of the relationship between microbial numbers, the microbial ecology of the food system and its spoilage mechanism continues to improve.

Models for predicting shelf life of foods that undergo non-microbiological deterioration (e.g. moisture- and oxygen-related changes) have also been developed and published (Floros & Gnanasekharan, 1993; McMurrough *et al.*, 1999). Most of them, however, are product-specific and require prior knowledge of some critical level of moisture, oxygen or other factor that causes the product to become unacceptable. As in the case of microbiological models, the complexity of foods makes it crucial to validate models using experimentally determined data in order to ensure appropriate and accurate prediction of shelf life.

1.12 What is challenge testing?

A challenge test is a laboratory investigation of the behaviour of a product when subjected to a set of controlled experimental conditions. In the context of shelf life determination, challenge testing refers to microbiological challenge testing, the aim of which is to simulate what can happen to a food product during processing, distribution and subsequent handling, following inoculation with one or more relevant microorganisms. The origin of microbiological challenge testing is believed to have come from the inoculated pack studies carried out in the early days of the canning industry. In these studies a highly heat resistant spore suspension of *Clostridium sporogenes*, a known spoilage organism, was used to challenge a processing system to determine the processing conditions which would reduce possible contamination with *Clostridium botulinum* to acceptable limits. A well known example of microbiological challenge testing is microbiological composition analysis (MCA) of edible emulsions developed by Tuynenburg Muys at the Unilever Research Laboratory in The Netherlands in the 1960s (Tuynenburg Muys, 1965, 1971). MCA has since been developed into the code for the production of microbiologically safe and stable emulsified and non-emulsified sauces containing acetic acid, commonly called the CIMSCEE Code (CIMSCEE, 1991). The code has two main parts: the first consists of formulae for predicting if a product is safe or stable at ambient temperature based on product composition (see Appendix B); the second consists of protocols for challenge testing products to establish safety and stability (Jones, 2000a). Clearly, challenge testing is a specialized laboratory exercise that is expensive, time-consuming and demanding on facilities and skills. Moreover, when a product formulation or the time–temperature profile to which it is subjected changes, challenge tests must be repeated. The main areas of application of microbiological challenge testing include:

- Determining product safety and assessing the risk of food poisoning after HACCP has identified the organisms likely to be a hazard for the product at some stage during production and distribution.
- Establishing shelf life by inoculating the product with food spoilage organisms likely to contaminate it.

- Evaluating the effects of different formulations of the food on a target organism, i.e. either a pathogen or a spoilage organism.
- Validating thermal processes such as aseptic processing and packaging, the effectiveness of which are expected to be very high and cannot therefore be established by monitoring failure rate during ordinary operations.

In all cases, relevant expertise and the necessary laboratory facility must be available to produce meaningful results. Detailed guidelines for the design and planning of microbiological challenge testing have been published (Rose, 1987; Notermans *et al.*, 1993). The increasing use of validated predictive models is expected to minimize the amount of challenge testing needed to establish microbiological safety and stability in the future.

1.13 Can the shelf life of my product be extended?

In many cases, the shelf life of food can be extended. The methods of shelf life extension must be founded on our understanding of the various mechanisms of food deterioration. From a purely scientific standpoint, our ability to extend the shelf life of a food product reflects our increasing understanding of its mechanism(s) of deterioration. A final decision to extend the shelf life of a food product is almost always a commercial one. It is pointless, for instance, to significantly increase the shelf life of a chilled food only to destroy its image of 'freshness' as a result. In practice, however, shelf life extension that brings about the following benefits is often welcome:

- Smoothing out production peaks and troughs.
- Offering wider choice to consumers.
- Stockpiling for seasonal increase in sale or special promotions.
- Widening of distribution.
- Less product wastage from actual product failure or insufficient time on the retail shelf.

Thus, establishing the main mechanism(s) of spoilage of a food product is the first step towards extending its shelf life. The next step is to see if the current shelf life can be extended simply by doing things better; this is a case of optimization. It may mean repeating the original storage trial, revisiting the major spoilage mechanism and re-examining the factors that contribute to it. Alternatively, new technology may have to be used in an attempt to extend shelf life. While there are different techniques of food preservation and extension of shelf life, the overriding objective in all these is always to minimize the occurrence and growth of microorganisms, although other non-microbiological forms of spoilage are usually controlled to varying degrees at the same time. A good understanding of the various preservation techniques that confer microbiological safety and stability to food can also aid the selection of the most appropriate method for shelf life extension. Principally, the major preservation techniques act by the following mechanisms (Gould, 1996):

- Inactivating microorganisms, e.g. pasteurization, sterilization, irradiation, high pressure processing.
- Preventing or inhibiting microbial growth, e.g. chilling, freezing, drying, curing, conserving, vacuum packaging, MAP, acidifying, fermenting, adding preservatives.
- Restricting the access of microorganisms to products, e.g. aseptic processing, decontamination (of raw materials, plant and environment), packaging.

In practice, these techniques can be used independently, or more commonly, in combination. A more recent trend is towards the use of procedures that deliver food products that are less severely preserved, without compromising safety, and often of higher quality, both real and perceived. Such procedures that make use of preservation factors acting in concert to give less damage to product quality have been called hurdle technologies (Leistner, 2000). Figure 1.3 illustrates the 'hurdle effect' using eight examples (Leistner, 1992).

Table 1.3 gives some examples of successful shelf life extensions that serve as evidence of our understanding of the deterioration mechanism involved in each case. In some cases (e.g. in commercial production of jam, sauces, salad dressings, ham, sausage and avocado purée), besides shelf life extension, the employment of a modern preservation technology such as high pressure processing

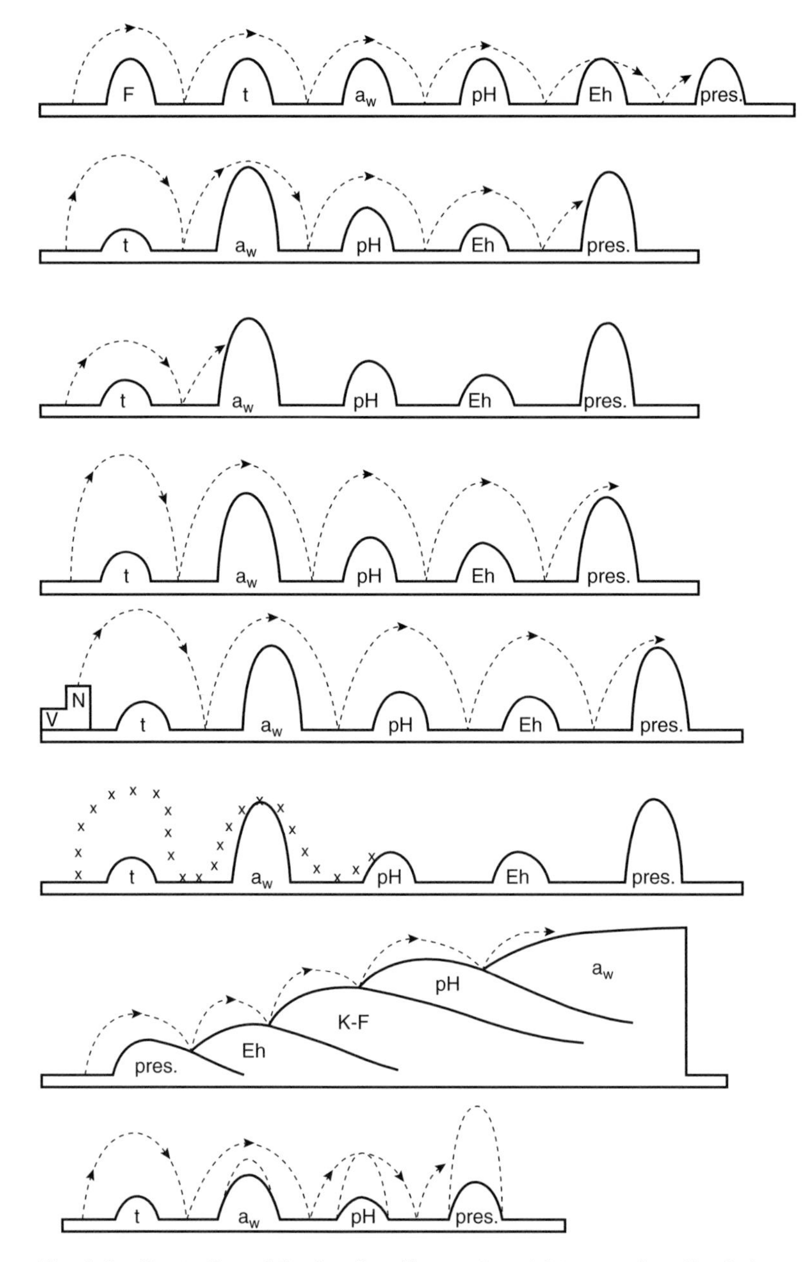

Fig. 1.3 Illustration of the hurdle effect, using eight examples. Symbols have the following meaning: F, heating; t, chilling; a_w, water activity; pH, acidification; E_h, redox potential; pres., preservatives; K-F, competitive flora; V, vitamins; N, nutrients.
(Reproduced from Leistner (1992) with permission from Elsevier Science.)

Table 1.3 Examples of successful shelf life extensions.

Food product	Main spoilage mechanisms	Technique of shelf life extension
Chilled foods, e.g. ready-to-eat sliced ham	Microbiological changes	Chilling and refrigerated storage Modified atmosphere packaging
	Biochemical changes	Use of oxygen scavengers
Refrigerated processed foods of extended durability (REPFEDs) – sous vide products	Microbiological changes	Vacuum packing Low temperature processing (65–95°C) Chill storage (0–3°C)
Sliced bread	Mould growth Staling Moisture transfer – redistribution and loss	Use of preservatives Use of emulsifiers Use of barrier packaging
Milk	Microbiological changes	Pasteurization Microfiltration (e.g. PurFiltre™)
Large fruit pies	Mould growth	UV irradiation
Orange juice	Microbiological changes Biochemical changes	HIgh pressure processing (Mermelstein, 1998)

has resulted in superior product quality compared with conventionally processed products (Johnston, 1994; Sizer, 2000). A concept that is developing into an important shelf-life extension technology is active packaging (Vermeiren *et al.*, 1999). The use of oxygen scavenging technology combined with MAP has extended the refrigerated storage lives of ready-to-eat red meat products. In other cases, even if shelf life extension is not appropriate or necessary, better understanding of food deterioration mechanisms should lead to improved assurance of and greater confidence in the established shelf lives of foods.

1.14 How are storage tests and trials set up for determining shelf life?

The most common and direct way of determining shelf life is to conduct storage trials of the product in question under conditions that mimic those it is likely to experience during storage, distribution, retail display and consumer use. This direct approach may be unacceptable if the expected shelf life is very long. In this case, alternative approaches such as ASLD have to be used (see Section 1.7). Of course, if the product being studied is a variant of established lines, an educated guess based on in-house technical expertise and sound scientific judgement is often sufficient for arriving at an estimate of its shelf life. The following aspects of direct storage trials deserve careful considerations.

1.14.1 Objective of the storage trial

The objective of the storage trial is a prime factor that determines how the experiment should be designed, planned and undertaken, and how the results should be interpreted. The same food destined for both retail sale and food service from a delicatessen counter where portions of the food are expected to be sold over a period of time, would require two different experimental designs to reflect the two applications.

1.14.2 Storage conditions

Storage conditions may be fixed or fluctuating. The actual storage conditions used will depend on the product being investigated and the amount of knowledge the experimenter has about the anticipated distribution chain through to consumer storage and use. Ideally, for a given set of storage conditions, the following variations should be available:

- Optimum conditions. They are the most desirable conditions of temperature, humidity, light and so on. Storage under these conditions should provide the most optimistic shelf life data.
- Typical or average conditions. They are the conditions most commonly experienced by the product. Storage under these conditions should provide shelf life data that apply to the bulk of future production most of the time.
- Worst case conditions. They are the most extreme conditions

that the product is likely to encounter. Storage under these conditions should provide the most conservative shelf life data which, if used to assign a shelf life, should give it a margin of safety ensuring that product failures due to insufficient shelf life are highly unlikely in practice.

Fixed storage conditions that are commonly used include:

- Frozen: −18°C or lower (relative humidity is usually near 100%).
- Chilled: 0 to +5°C, with a maximum of +8°C (relative humidity is usually very high) (HMSO, 1995b).
- Temperate: 25°C, 75% relative humidity (Cairns, 1974).
- Tropical: 38°C, 90% relative humidity (Cairns, 1974).
- Control: control conditions (for storage of control samples) are usually the optimum conditions, be they ambient, chilled or frozen.

Different countries, even within the EU, may have different requirements. For instance, Belgium (and The Netherlands) and Spain, respectively, stipulate a maximum of 7°C and 0–3°C for the storage of chilled foods (Goodburn, 2000). For chilled foods destined for exports, the storage conditions stipulated by the country of destination must be used for storage trials. Storage under fluctuating conditions makes use of a programmed storage facility that creates a set of artificial conditions (e.g. heating or lighting coming on and off according to a predetermined pattern) that are designed to mimic the real-life conditions experienced by the product. Such a facility is obviously expensive and so fixed conditions storage tends to be the preferred storage for the direct determination of shelf life. Whatever the conditions, they must be closely monitored and recorded to ensure correct and proper interpretation of shelf life data.

1.14.3 Samples for storage trials

The product composition or formulation, the way the samples have been produced as well as the packaging materials used are important factors; they need to be noted and controlled. For instance, pilot-scale samples are likely to have been produced on a batch basis, whereas production-scale samples are more likely to have been processed on a semi- or fully continuous basis. The differences in product characteristics between pilot-scale and production-scale samples may well be enough to have significant bearing on the outcome of the storage trials.

The number and the size of the samples need to be carefully chosen, consistent with the objective of the storage trial. Ideally, the food should be stored in the same pack or container that has been designed and developed for full-scale production. Care must be taken to ensure that all sample packs are exposed to exactly the same storage conditions.

The number of samples to be taken is very much dictated by the sampling schedule for the storage trial. In turn, the sampling schedule is influenced by the type of product, its end-use application, the anticipated or required shelf life and the tests to be carried out for assessing changes during storage. In one example, ambient shelf-stable pasta shapes in savoury tomato sauce packed in multilayered plastic trays with a desired shelf life of 1 year was studied. A total of some 350 samples were required for storage trials at three different temperatures, i.e. 2°C (control), 25°C (normal) and 35°C (worst case) (Goddard, 2000).

Not all types of product are unaffected by freezing and thawing. When frozen storage is unsuitable as a means of keeping control samples, facilities must be available for the preparation of fresh reference (i.e. control) samples that are identical to the test samples in every way, at any time during a storage trial.

1.14.4 *Sampling schedule*

Different designs of shelf life experiment based on a statistical approach have been published (Gacula, 1975). In practice, however, the actual sampling schedule chosen is often determined by the shelf life anticipated by the experimenter or the shelf life required by the customer. As an illustration, the following are some possible sampling schedules:

- Short shelf life products. For chilled foods with shelf life of up to 1 week (e.g. ready meals), samples can be taken off daily for evaluation.
- Medium shelf life products. For products with a shelf life of up to 3 weeks (e.g. some ambient cakes and pastry), samples can be taken off on days 0, 7, 14, 19, 21 and 25.
- Long shelf-life products. For products with a shelf life of up to 1 year (e.g. some breakfast cereals and heat-processed shelf-stable foods), samples can be taken off at monthly intervals or at months 0, 1, 2, 3, 6, 12 and (perhaps) 18. The exact frequency will depend on the product and on how much is already known of its storage behaviour.

1.14.5 Shelf life tests

The exact shelf life tests are often product-specific and may include some or all of the following types of tests (see also Section 1.9):

- Microbiological examination including challenge testing.
- Chemical analysis.
- Physical testing, measurement and analysis such as rheological measurements, microscopical examination, vibration test, and so on.

And in all cases:

- Sensory evaluation.

Given the assurance of product safety, sensory evaluation is undoubtedly the most appropriate type of test for evaluating changes during storage trials. To ensure the generation of meaningful, accurate and reliable sensory data, some basic and inter-related requirements have to be fulfilled; they are as follows (Kilcast, 2000):

1. Objectives of the sensory evaluation must be clearly defined.
2. A dedicated sensory testing environment must be available.
3. Suitable test procedures must be used.
   - Analytical tests (product-oriented tests): difference

(discrimination) tests and quantitative tests, e.g. quantitative descriptive analysis (QDA).
- Hedonic tests (consumer-oriented tests): preference and acceptability tests.

4. Suitable assessors (i.e. taste panellists) must be selected and trained.
5. Data handling and analysis must be correct and the results presented effectively.

Detailed discussions and guidance on the use of sensory evaluation in shelf life testing can be found in a number of publications (IFT, 1981; O'Mahony, 1986; Labuza & Schmidl, 1988; Stone & Sidel, 1993; Lawless & Heymann, 1998; Carpenter *et al.*, 2000).

1.15 Summary

The following are a number of the key points:

- Shelf life is an important requirement of today's food products.
- Safety and consistent quality that meet customer expectations are the two main aspects of an acceptable shelf life.
- Within the EU and in the UK, the provision of an appropriate and reliable date of minimum durability on food labels is a legal requirement; this date depends on the food's shelf life.
- The responsibility of determining shelf life of a food product lies with its manufacturer and/or packer.
- Management understanding and commitment are essential if shelf life determination is to be taken seriously, because significant resources are needed to do the work properly.
- Shelf life is determined directly by conducting storage trials of the product under defined storage conditions.
- In many cases, and for a number of reasons, shelf life may be estimated, predicted or determined indirectly by accelerated tests, microbiological challenge tests and/or the use of suitable computer programs.
- Knowledge of the relevant spoilage mechanism(s) of a food

product is crucial to its shelf life determination, and if necessary, its shelf life extension as well.

- Shelf life of foods is rarely affected by a single factor; a number of factors influencing shelf life are usually at work.
- The most effective way of managing shelf life is the careful application of GMP principles in food manufacture and processing.

Section 2

The Ways Food Deteriorates and Spoils

2.1 Mechanisms of food deterioration and spoilage

All foods deteriorate. The ways by which they deteriorate and spoil can be complex; often more than one mechanism may be taking place at the same time. Knowing and understanding the mechanism(s) of deterioration of a food product enables the product developer or shelf-life investigator to identify and then control the factors that have a major influence on its shelf life. While it may not be feasible to prevent the deterioration from occurring altogether, solutions may be found to delay it or minimize the extent of its impact so that a commercially acceptable shelf life can be assigned to the product. For the sake of simplicity, the most well known mechanisms are considered in the following sections.

2.1.1 Moisture and/or water vapour transfer

In many food products, water is often a major as well as one of the most important components. Not only is water a medium for chemical and biochemical reactions, but also it can participate in some of them. From a microbiological point of view, water is one of the most critical factors, which determines whether or not microbial growth will occur. Equally important is the part water plays in affecting the sensory properties of foods. Consequently, many food products are sensitive to a gain or loss of moisture (or water vapour), depending on the direction of moisture and/or water vapour transfer. Table 2.1 gives some examples of this deterioration.

Table 2.1 Examples of quality changes due to moisture and/or water vapour transfer.

Product	Quality change	Deterioration mechanism
Fresh vegetables	Wilting	Moisture loss
Fresh soft fruit	Dry, unattractive appearance	Moisture loss
Prepared ready-to-eat salads	Loss of gloss, crispness	Moisture loss
Dressed salads, e.g. coleslaw	Changes in texture of vegetables, changes in consistency of dressing	Moisture migration from vegetables to dressing
Biscuits	Softening, loss of crunchiness	Moisture gain
Breakfast cereal flakes	Loss of crispness, crunchiness	Moisture gain
Ambient packaged cakes	Hardening of texture, drying	Moisture loss
Savoury snacks	Loss of crispness	Moisture gain
Hard-boiled sweets	Increased stickiness	Moisture gain
Powder beverages	Caking, reduced wettability	Moisture gain
Frozen meat	Freezer burn	Water vapour transfer – ice sublimation

Transfer of moisture or water vapour will take place between adjacent components in a composite product as long as a gradient exists; the magnitude of the gradient will have a major influence on the rate of transfer. Also, moisture transfer will not stop at sub-zero freezer temperatures, so it is now a standard practice to coat ice cream wafer cones internally with a layer of fat-based coating, e.g. a simulated chocolate coating that helps to delay the softening effect of moisture migration on the cones. Protection will only be complete, however, if the ice cream cones are packaged in moisture barrier materials that protect them from the atmosphere of the display cabinet.

2.1.2 Physical transfer of substances other than moisture and/or water vapour

The transfer either into or out of food, of substances other than moisture which affect its safety and/or quality, is likely also to have an impact on its shelf life. The following are some examples of quality changes due to this mechanism of quality deterioration.

Quality change due to transfer out of food:

- Loss of carbon dioxide from carbonated drinks packaged in PET bottles, resulting in loss of fizziness; as much as 60% of the starting level of carbonation can be lost from 500 ml PET bottles of tonic water and lemonade after only 6 months' storage at ambient temperature (Matthews, 1995).
- Sorption of limonene and other aroma compounds (i.e. flavour scalping) by packaging material from orange juice packaged in high-density polyethylene bottles, resulting in possible reduction in citrus flavour intensity (Nielsen & Jägerstad, 1994).

Quality change due to transfer into food:

- Taint and off-flavour development as the food picks up foreign and objectionable flavours and odours, depending on the packaging used and the prevailing environment; foods that are susceptible to taint include leaf tea (large surface area) and chocolate confectionery (high fat content).
- Substances such as monomers and additives (e.g. in plastics), and heavy metals (e.g. in paper boards (Conti, 1998)) can migrate from packaging materials into foods, potentially giving rise to safety and/or quality problems. In the EU, migration from food contact materials is regulated by the framework Directive 89/109/EEC (EC, 1989) relating to materials and articles intended to come into contact with foodstuffs.

2.1.3 Chemical and/or biochemical changes

Food is composed of chemicals and most of the raw materials used in the manufacture of food products are biological in origin. Consequently, some chemical and biochemical changes in foods are inevitable. Apart from a few changes such as maturing of cheeses or wines and post-harvest ripening of fruits, most chemical and biochemical changes in food are undesirable and therefore shelf life limiting. These major chemical and biochemical changes are oxidation, hydrolysis, non-enzymic browning, enzymic browning and interactions between food and packaging.

Oxidation

Oxidation of fats and oils

Oxidation of fats and oils leads to the development of undesirable 'off-flavours' (rancidity) and odours, and hence reduced consumer acceptance or even rejection. Oxidative rancidity (sometimes referred to as autoxidation) is a chemical reaction with low activation energies (4–5 kcal mol^{-1} and 6–14 kcal mol^{-1} for the first and second steps, respectively) and is therefore not stopped by lowering the temperature of food storage. It can occur by one of two mechanisms: the classical free radical mechanism involving a catalyst, e.g. copper ions, which can operate in the dark, or a photo-oxidation mechanism in the presence of a sensitizer, e.g. myoglobin, which is initiated by exposure to light (Hamilton, 1994). The first product in both cases is a lipid hydroperoxide, which is itself odourless, but which breaks down to smaller molecules that produce the rancidity. There is yet a third mechanism of oxidation, which is the lipoxygenase route. The enzyme lipoxygenase is believed to be widely distributed in foods from both the plant and animal kingdoms, e.g. cereals (wheat, barley, maize), oilseeds (soya, groundnut), and fish. The oxidative deterioration again involves a hydroperoxide intermediate that eventually gives rise to off-flavour products.

Oxidative rancidity decreases the nutritional quality of food because the free radicals and peroxides generated destroy polyunsaturated fatty acids and fat-soluble vitamins A and E in food. These intermediates can also react with sulphydryl bonds in proteins. As sulphur amino acids are often the nutritionally limiting amino acids in many proteins, a reduction in their content will invariably lead to a decrease in protein quality. Several classes of material in oxidized fat are known to have toxic effects. They include peroxide fatty acids and their subsequent end-products, polymeric material and oxidized sterols. Polycyclic aromatic hydrocarbons produced from the pyrolysis of fats on grilling and roasting meat and fish are known carcinogens (Sanders, 1994).

Oxidation of food pigments

All natural pigments are unstable. While loss of or change in the natural colour in food does not necessarily mean a reduction in its nutritive value, the contribution made by colour to overall appearance of food and hence its acceptability is vitally important. A good example is the colour of fresh meat, which is due to myoglobin that can exist in three forms, namely red oxymyoglobin, purple reduced myoglobin and brown metmyoglobin. Discolouration of meat is caused by the oxidation of oxymyoglobin and myoglobin to metmyoglobin. Consequently, colour stability of red meat is often taken as an indicator of its shelf life as well as its freshness.

Oxidation of vitamins

Vitamins are one of the few groups of food components in which it is possible to demonstrate quantitatively a reduction in quantity over a period of time (Berry-Ottaway, 1993). Chemically, they are a heterogeneous group of compounds with no common structural features or deterioration mechanism. Nevertheless, a number of them are sensitive to oxygen (e.g. water-soluble: vitamin C (L-ascorbic acid) and vitamin B_1 (thiamine); fat-soluble: vitamin A and vitamin E). With the increased use of vitamins in foods such as fortified breakfast cereals and functional sports drinks, the levels of vitamins declared on the labels as well as in the unopened packs at the point of sale can be conveniently used as a shelf life end-point for these products (see also Section 1.9). To meet label claims and to cater for vitamin degradation during the expected shelf life, manufacturers usually add more than the amounts stated on the label. The difference between formulated and declared levels is known as the 'overage' (Berry-Ottaway, 1993). Overages are normally expressed as a percentage of the declared level:

$$\frac{\text{Amount of vitamin present in product} - \text{Amount declared on product label}}{\text{Amount declared on product label}} \times 100\%$$

Overages to be used will depend on the inherent stability of the vitamins, the type of products (i.e. fortified or enriched), the conditions under which the food is manufactured and distributed, and the target shelf life. They are usually estimated by the food manufacturer during product development, based on past experience of similar products and/or storage trials at different temperatures and analysis of the data obtained using the Arrhenius model. Examples of vitamin overages that have been used are given in Table 2.2. When more than one vitamin is added, their stability can be more difficult to estimate, as has been shown to be the case, at least in model food systems (Pacquette, 1998). Where appropriate, the exclusion of oxygen (and in some cases, light as well) by the use of suitable packaging materials remains an effective means of protecting vitamins in food, both added and indigenous, from early degradation.

Hydrolysis

This is the splitting of molecules, under suitable conditions, in the presence of water. In certain foods, some hydrolytic reactions can be shelf life limiting.

Table 2.2 Examples of vitamin overages used in some products[a].

	Product (required shelf life)	
Overages (%)	Milk based fortified drink powder (12 months at ambient)	Fortified meal replacement bar (12 months at ambient)
Vitamin A	25	45
Vitamin D	25	30
Vitamin E	10	10
Vitamin B_1	15	15
Vitamin B_2	15	20
Niacin	10	15
Vitamin B_6	20	30
Pantothenic acid	15	30
Folic acid	20	25
Biotin	20	20
Vitamin B_{12}	20	20
Vitamin C	30	35

[a] Adapted from Berry-Ottaway (1993).

Hydrolysis of aspartame

Aspartame is a high-intensity sweetener commonly used in 'diet' soft drinks and other low energy ('calorie-free') products. Under the right conditions of temperature and pH, it will hydrolyse slowly resulting in a gradual reduction in sweetness of the product. This gradual loss in sweetness can become a shelf life limiting factor.

Hydrolysis of oils and fats

Hydrolysis of the triglycerides (oils and fats) in the presence of moisture and an enzyme, liberates short-chain (C_6–C_{10}) free fatty acids that have strong off-flavours (i.e. rancidity). The enzyme involved is usually a free lipase or esterase. Hydrolytic rancidity is mainly encountered in products based on lauric oils such as palm kernel or coconut. The free fatty acids liberated comprise large amounts of capric, lauric and myristic acids, which have a distinct soapy flavour; hydrolytic rancidity is therefore often referred to as soapy rancidity. Lipases are also found in various cereal grains and milling products, e.g. wholegrain wheat, wheat bran, oats, brown rice and rice bran. Heat treatment is common practice to inactivate the enzyme and stabilize these products which, unfortunately, is not appropriate for brown and wholemeal wheat. When rancidity develops in dairy products, it is usually either due to the use of poor quality milk or microbial contamination of the product during or after processing. Hydrolytic changes initiated by enzymes may occur in meat or meat fats where there is microbial growth but they are not very common in ordinary meat.

One of the few cases where hydrolytic rancidity is regarded as desirable is in some strong cheeses such as Stilton, where it gives the cheese a sharp burning taste.

Non-enzymic browning

The Maillard reaction is a well-known mechanism of non-enzymic browning. It involves the reaction of an aldehyde (usually a reducing sugar) and an amine (usually a protein or amino acid). The rate of this reaction increases with increases in temperature, heating time, and free amine and aldehyde groups. The reaction is also promoted by alkaline conditions and enhanced by phosphates (Bell, 1997). It is characterized by browning of products accompanied by a loss of nutritive value, principally from losses in lysine, an essential amino acid that reacts readily with reducing sugars. The antimicrobial (Painter, 1998) and carcinogenic (Skog *et al.*, 1998) effects of Maillard reaction products are also well established. Products such as dehydrated fruits and vegetables, instant potato powder, dried egg white and dried milk products are known to be susceptible to the Maillard reaction which can limit their shelf lives.

Sulphite additives are commonly used to control non-enzymic browning in foods, but while vitamin C is a powerful preservative against enzymic browning, its degradation has been shown to be a major contributor to non-enzymic browning in products such as aseptically produced orange juice (Roig *et al.*, 1999).

Deep fat frying also results in browning. Lipid browning is perhaps the least well studied of the non-enzymic browning reactions. It is believed to be the polymerization or reaction of carbonylic oxidation products of unsaturated lipids with amino compounds.

Enzymic browning

Browning in some fruit (e.g. peeled apples, bananas) and vegetables (e.g. cut lettuce leaves and mushroom caps), is caused by the enzymic oxidation of phenolic compounds (i.e. the substrate) initiated by the enzyme polyphenol oxidase, which is also known as phenolase. Such enzymic browning is seen by most consumers as undesirable and it can limit shelf life in products such as pre-cut lettuces and fresh fruit salads. The most important factors that determine the rate of enzymic browning of fruit and vegetables are the concentrations of both active phenolase and phenolic compounds present, the pH, the temperature and the oxygen availability of the tissue (Martinez & Whitaker, 1995). The control of enzymic browning as a means of extending shelf life currently relies on both physical and chemical methods, often used together. The best control procedures, however, differ between foods as there are few generally applicable measures. For instance, heat treatment such as pasteurization, which is effective in denaturing the enzyme and so halting browning, cannot be applied to fresh fruit and vegetable salads. However, pineapples have been successfully packed in a CO_2-enriched atmosphere, limiting the availability of the oxygen substrate. A very successful and widely used method of controlling enzymic browning has been the use of anti-browning sulphite additives such as sodium metabisulphite. Sulphite additives, however, have come under scrutiny because of their possible harmful side-effects in humans. To date, the most effective and safest inhibitors of enzymic browning are L-ascorbic acid (vitamin C) and its isomers and derivatives. A useful review of new approaches to controlling enzymic browning in foods is available (Walker & Ferrar, 1995).

Enzymic browning is sometimes desirable, as it can improve the sensory properties of some products such as dark raisins and fermented tea leaves.

Interactions between food and packaging

One of the interactions between packaging and food that has been studied in depth is that between foods and tinplate cans. Here, the interactions are electrochemical reactions involving electrodes (container and lid or closure) and electrolytes (food products). The precise reaction that occurs will be influenced by a number of factors such as the number and type of metals present, the type of food and the presence or absence of air within the pack (Turner, 1998). Thus, although tin dissolution from tinplate surfaces is generally to be avoided, it is essential for good pack performance for some classes of foods such as white fruits that are subject to oxidative discoloration and off-flavour (taint) development. In normal plain tinplate containers, the sacrificial detinning of the container in a number of acid products ensures the rapid consumption of any residual oxygen in the pack, leading to more controlled and much reduced detinning, and maintenance of the assigned shelf life. In effect, the tin gets oxidized, protecting the steel underneath it. The dissolved tin also imparts a characteristic 'bite' to products such as citrus juices and tomato-based products (e.g. baked beans in tomato sauce). The colour of a product, too, may also be favoured by the reducing action of tin, so that mushrooms, asparagus, potatoes and carrots, for instance, have traditionally been packed in plain tinplate containers. Products that contain anthocyanin pigments (e.g. cherry, strawberry), however, are packed in lacquered containers because dissolution of tin causes discoloration and results in commercially unacceptable products due to the reduction of the pigments. In the UK, the amount of tin in foods is regulated by the *Tin in Food Regulations* (HMSO, 1992), which prohibit the sale or importation of any food containing a level of tin exceeding 200 milligrams per kilogram (see also Section 1.9).

In the other direction, there are two broad types of attack by the product on the container (Turner, 1998):

- Acid attack with evolution of hydrogen, solution of metal ions and, in extreme cases, perforation of the container; this is normally associated with acidic products.
- Conversion of the metal surface by ingredients of the product, typically the formation of iron and tin sulphides (i.e. iron sulphide and tin sulphide staining) resulting from the interaction of the metal surface and sulphur compounds derived from the degradation of protein during the retorting process. Table 2.3 gives some examples of general specifications used for various classes of food products in Europe and the USA (Turner, 1998).

2.1.4 Light-induced changes

The following factors are important when considering the mechanism of deterioration induced by light (IFST, 1993):

- The wavelength of the light (e.g. UV or visible).
- The intensity.
- The duration of exposure.
- The absorbability of the food.
- The presence of sensitizers.
- The environmental temperature.
- The amount of available oxygen.

The actual change depends on the food. The following light-induced or accelerated changes are well known:

- Photo-oxidation of some vitamins. Ascorbic acid, for example, decomposes rapidly in the presence of oxygen and light. Decomposition of the vitamin increases with increasing light intensity as long as oxygen is not limiting, and is also accel-

Table 2.3 Examples of container specifications for various food products in the UK – welded three-piece bodies[a].

Products	Specification
White fruits: apples, pears, pineapples, apricots, grapefruit Oranges/orange juice, tomatoes Asparagus, baked beans, carrots, celery, potatoes, spinach	Plain cans (high tin-coating, e.g. $11.2\,g\,m^{-2}$)
Nut-based products, fruit mincemeat, lemon curd Green beans, coleslaw, mixed vegetables, pulses, garden and processed peas	Epoxy-phenolic lacquer – single coat
Crab meat, fish in oil	Epoxy-phenolic lacquer – single coat unpigmented
Brussel sprouts, sweetcorn	Epoxy-phenolic lacquer – single coat with zinc pigmentation
Meat and fish: shellfish in brine (e.g. cockles and mussels); sliced meats, paté, poultry, stews and casseroles, meat pie fillings	Epoxy-phenolic lacquer – single coat with zinc pigmentation, or other sulphur-absorbing or masking pigment
Red fruits: blackberries, blackcurrants, red cherries, red plums, damsons Grape juice, coloured fruit pie fillings Pickled gherkins, olives, onions Shellfish in vinegar	Epoxy-phenolic lacquer – two coats
Rhubarb	Two- or three-coat combinations of epoxy-phenolic and organosol lacquers
Solid meat packs: corned beef, hams, luncheon meat, burgers, paté	Epoxy-phenolic lacquer – single coat meat-release lacquer

[a] Adapted from Turner (1998).

erated by riboflavin which acts as a photosensitizer (Jung *et al.*, 1995).

- Light-induced oxidation of milk. Although extremely rare today, this can easily be induced in fresh milk left on sunny doorsteps, giving rise to an off-flavour described as 'cardboard-like' due to oxidation of the fat in milk.
- Photo-oxidation of nitric oxide pigments in cooked ham products. The colour of nitrite cured meat products is due to nitric oxide pigments which inhibit lipid oxidation (Kanner *et al.*,

1980, 1984). In the presence of even minimal amounts of oxygen, light induces photo-oxidation of the nitric oxide pigments, causing severe discoloration of the products. It has been demonstrated that interactive packaging using oxygen absorbers protects these products against photodegradation of their colour (Andersen & Rasmussen, 1992) (see also Section 1.13).

- Light-accelerated development of oxidative rancidity in foods. Prolonged exposure to artificial light (e.g. from fluorescent light tubes used in supermarkets) at ambient humidity can cause potato crisps to become rancid even before any textural changes due to moisture uptake are detected (Man, 2000). For this reason, most savoury snacks are packaged with a light barrier, e.g. a metallized film. The same principle has, for a long time, been applied successfully to the packaging of bakery products such as biscuits, to exclude light and minimize the ingress of oxygen.
- Photodecomposition of aspartame. Besides being a high-intensity sweetener, aspartame has flavour-enhancing properties. It has limited stability in aqueous solutions and its decomposition is significantly increased by light, the photodecomposition being proportional to the light intensity (Kim *et al.*, 1997). At pH 7, the photodecomposition is accelerated by riboflavin which acts as a sensitizer.
- Colour fading due to light sensitivity. Colours are added to foods mainly for decorative purposes, i.e. either to improve consumer appeal or to replace colours lost as a result of processing. Many nature identical (e.g. riboflavin (yellow)) or naturally derived colours (e.g. curcumin (yellow) and chlorophyll (green)) are light sensitive, causing colour fading. Their poor stability limits their use in frozen or short shelf life products. For other applications, the most common solution is to add a permitted artificial colour that is usually more stable and able to meet the shelf life requirement (Downham & Collins, 2000).

2.1.5 Microbiological changes

In principle, all foods, in particular moist foods, are ideal substrates for microbial growth which, if allowed to take place, will lead either to food poisoning or spoilage. Factors affecting microbial growth are as follows (Mossel, 1971):

- Intrinsic properties of the food (e.g. nutrients, pH, total acidity, water activity, structure, presence of preservatives and/or natural antimicrobials, redox potential).
- Extrinsic factors (e.g. environmental temperature, relative humidity, gaseous atmosphere).
- Processing factors (e.g. heat destruction, freezing, packaging)
- Implicit factors (e.g. physiological attributes such as specific growth rate of the microorganisms and microbial interactions).

Biological and microbiological foodborne diseases

Most foodborne diseases are caused by bacteria that initially contaminate the living plant or animal or recontaminate the food during handling after processing. Some of the well known foodborne diseases are given in Table 2.4. Besides the microbiological hazards in foods caused by pathogenic bacteria there are other biological

Table 2.4 Causes of foodborne diseases (Sinell, 1995).

'Classic'		Emerging	
Infections	Intoxications	Infections	Intoxications
Salmonella spp. *Clostridium perfringens* *Vibrio parahaemolyticus* *Escherichia coli* Milk borne, e.g. *Mycobacterium tuberculosis* *Shigella* spp.	*Clostridium botulinum* *Staphylococcus aureus* *Bacillus cereus*	*Campylobacter* spp. *Yersinia* spp. *Escherichia coli* 0157 (verocytotoxic) *Listeria monocytogenes* *Aeromonas* spp. *Plesiomonas* spp. *Edwardsiella* spp. *Salmonella enteritidis* *Anisakis* spp. *Pseudoterranova* spp.	Paralytic shellfish poisoning (PSP) Diarrhetic shellfish poisoning (DSP) Histamine poisoning (scombrotoxicity)

hazards that must not be overlooked and they are the following (Untermann, 1998):

- Macroparasites, e.g. *Trichinella spiralis*, trematodes.
- Protozoae, e.g. *Sarcocystis* spp., *Giardia lamblia, Cryptosporidium parvum* (IFST, 1997a).
- Mycotoxin-producing fungi, e.g. *Aspergillus flavus.*
- Viruses (IFST, 1997b), e.g. *Hepatovirus* (hepatitis A), small round structured virus (SRSV).
- Prions, e.g. bovine spongiform encephalopathy (BSE)?

The general view held by most experts is that foodborne diseases are preventable. The first priority of every shelf life study must therefore be to ensure the safety of the food being evaluated, particularly from a biological and microbiological viewpoint. Published information is available to assist with the hazard analysis of food pathogens (ICMSF, 1996). In future, successful development of an international microbiological hazard database will further facilitate hazard identification and ensure correct implementation of HACCP (Panisello & Quantick, 1998). Special consideration must be given to the target population at which the product whose shelf life is being studied is aimed. Vulnerable consumer groups or sensitive populations who may be at greater risk of serious illness are now believed to represent around 25% of the UK population. These groups include (Gerba *et al.*, 1996; IFST, 1998):

1. Children under the age of five.
2. Pregnant women.
3. The elderly (over 65).
4. Residents in nursing homes or related care facilities.
5. Persons who have an impaired immune system such as
    - people taking immunosuppressive drugs (e.g. transplant organ patients),
    - patients undergoing cancer therapy,
    - people infected with the HIV virus, which causes AIDS.

Microbiological spoilage

Over the years, the microbiology of food spoilage has received considerable attention. The characterization of the typical microflora that are associated with different types of foods during storage has been well documented (Mossel *et al.*, 1995; ICMSF, 1998). In certain foods such as seafood, the successful characterization of the part of the total microflora responsible for spoilage has led to the development of the specific spoilage organisms (SSOs) concept (Dalgaard, 2000). This concept forms the basis of the specific spoilage models within the SSP program (see Section 1.11). As yet, SSOs have not been determined in other products and the general application of this simple concept has not been fully established.

For convenience, food spoilage organisms may be broadly divided into the following groups (Huis in't Veld, 1996):

- Gram-negative rod shaped bacteria. *Pseudomonas* spp. are among the most common spoilage organisms, particularly in aerobically stored foods with a high water content and high natural pH, e.g. red meat, fish, poultry, milk and dairy products. Spoilage is characterized by production of off-odours, visible slime and pigmented growth.
- Gram-positive spore forming bacteria. These include *Bacillus* and *Clostridium* spp. capable of surviving a thermal process. A well known member of the former is *Bacillus cereus*, which may grow at 5°C or lower and produce enzymes that cause 'sweet curdling' and 'bitty cream' in milk. In general, spoilage *Clostridium* spp. are unable to grow at 5°C or lower. At slightly higher temperatures, they may produce gas causing, for example, 'late blowing' of hard cheeses during maturation.
- Lactic acid bacteria. Typical members of this group are *Lactobacillus*, *Streptococcus*, *Leuconostoc* and *Pediococcus* spp. They spoil foods by fermenting sugars to form lactic acid, slime and carbon dioxide, causing a drop in pH and off-flavour development.
- Other Gram-positive bacteria. *Brochothrix thermosphacta* may occasionally be present on fresh meats. The increased use of modified atmosphere packaging and vacuum packaging has often allowed *Br. thermosphacta* to dominate the microflora, resulting in off-flavour development.
- Yeasts and moulds. They can be found in a wide variety of environments and can utilize a wide range of substrates. They are relatively tolerant to low pH, low water activity, low tem-

perature and the presence of certain preservatives. Consequently they can contaminate a wide range of foods and beverages, causing the production of soft rot (e.g. fruit), pigmented growth (e.g. baked goods), fermentation of sugars to give acid, gas or alcohol (e.g. soft drinks, jams) and development of off-odours (e.g. beer).

Clearly, microbiological spoilage and biochemical changes can be closely related and they may interact. The underlying integrated spoilage mechanisms, however, are still poorly understood.

2.2 Factors influencing the shelf life of foods

The actual factors that influence the shelf life of a food product and its mechanism(s) of spoilage are closely related. Knowledge of the mechanism of spoilage will aid identification of the correct shelf life limiting factors. Once identified, the latter and their associated parameters will need to be controlled and monitored, in much the same way as in HACCP, if the assigned shelf life is to be maintained consistently. In general, there are many factors that can influence shelf life; they are considered in the following sections.

2.2.1 Intrinsic factors

Raw materials

Almost as a rule, the quality of a finished product is a reflection of the quality of its raw materials. Not all the quality characteristics and parameters of a raw material will have an influence on shelf life. Those that do will need to be recognized and their effect on shelf life established. For example, cabbages from cold storage generally have a higher yeast count than freshly harvested ones (Betts & Everis, 2000). Use of the former in coleslaw is likely to result in a markedly reduced shelf life, requiring an amendment to the standard shelf life for parts of the year. As their quality is paramount, realistic purchasing specifications are needed for all raw materials.

Product composition and formulation

The composition of a food product can be the single most important shelf life determining factor in many products. For example, the high refractometric solids give traditional jams (i.e. with no added preservative) their long ambient shelf lives. Similarly, a preservation index of 3.6% acetic acid in the volatile constituents of non-pasteurized pickles and sauces confers upon them microbiological safety and stability. Fried foods contain a significant percentage of oil (Table 2.5) the quality of which is of paramount importance in determining their shelf lives. Margarine is a water-in-oil emulsion with a high fat content (minimum 80%) that limits the growth of most microorganisms, including pathogens and spoilage organisms (Delamarre & Batt, 1999; Tuynenburg Muys, 1969). Although rare today, it is still prone to oxidation. The development of low-fat spreads that have a reduced fat content, has seen a shift from oxidation as a potential deterioration mechanism to other mechanisms such as emulsion instability and microbiological spoilage. Thus emulsifiers (e.g. mono- and diglycerides and lecithin) and preservatives (e.g. potassium sorbate) are usually added to increase the shelf lives of these products. Research has demonstrated that, besides an adequate heat process and effective temperature control during subsequent storage and distribution, product formulation (in respect of pH, salt content, etc.) is a critical factor in determining the shelf lives of *sous vide* foods (Gould, 1999).

Food structure

Many solid and semisolid food products (e.g. sausages, mayonnaise, margarine) do not have a truly homogeneous and uniform structure.

Table 2.5 Oil absorption in fried foods – typical values.

Food	Oil absorbed (%)
Frozen chips	4–6
Fresh chips	9–11
Battered food (e.g. fish)	8–12
Battered and breaded food	7–13
Doughnuts	10–15
Potato crisps	30–35

Consequently, the chemical and physical conditions relevant to microbial growth and/or chemical and biochemical reactions can vary with position in the food microstructure. Electron microscopy studies (Katsaras & Leistner, 1991) have revealed that the natural flora and the added starter cultures are not evenly distributed in fermented sausages, but are arrested in cavities of the product. The distribution of the microbes has been shown to influence the ripening process and hence, the microbiological safety and stability of the product. Food structure, too, has helped to explain why predictive microbiological models sometimes fail when applied to structured foods (Robins *et al.*, 1994; Dens & Van Impe, 2001). Similarly, food structure is believed to account for the significant differences between lipid oxidation in bulk fat and that in emulsified fat (Coupland & McClements, 1996).

Product make-up

Product make-up or the assembly of product may be viewed as the macrostructure of food. In multicomponent and composite products, contact between components often result in migration of moisture, colours, flavourings or oil from one component to another. In fruit pies, migration of moisture from the filling to the pastry leads to a gradual loss of the desired texture. Moisture migration can occur in the vapour phase when cereal flakes are mixed with dried fruit, which usually has a shorter shelf life than the flakes alone (Howarth, 2000). In a similar way, combining components of different microbiological status such as diced cheese and coleslaw can result in a shorter shelf life for the end product.

Water activity value (a_w)

Water activity (a_w) expresses the 'availability' of the water in a given solution. When this solution and the atmosphere with which it is in contact are in equilibrium, the relative humidity of that atmosphere is called the equilibrium relative humidity (ERH). Under a defined set of conditions of atmospheric temperature and pressure, the relationship between these two measures is described by the following equations:

$$a_w = \text{ERH}/100$$
$$\text{ERH} = a_w \times 100\%$$

Water activity values have been widely used to indicate the stability of foods with respect to the potential for microbial growth, chemical and biochemical changes, and physical transfer such as moisture migration (Labuza & Hyman, 1998). Humectants such as invert sugar, glycerol, dextrose and various glucose syrups are used in food formulations to influence this potential. Water activity values of some foods and the minimal a_w values that support microbial growth of selected organisms are given in Tables 2.6 and 2.7. These values, however, should not be regarded as absolute values. More importantly, a_w is only one of a number of factors that can affect microbial growth (see Section 2.1.5) and it is the interplay between them that ultimately determines whether or not a microorganism will grow in food.

Table 2.6 Water activity (a_w) values of some foods.

a_w	Food
>0.98	Fresh meats, fish Fresh fruits, vegetables Milk, creams Fruit juices
0.98–0.93	Cooked sausages Some cheeses, e.g. Cheddar and processed cheese Cured meats, e.g. ham Evaporated milk Breads, crumpets
0.93–0.85	Dry or fermented sausages Dried beef Raw ham Aged Cheddar cheese Sweetened condensed milk
0.85–0.60	Flour confectionery products, e.g. ambient cakes and pastries Dried fruit Jam and jellies Heavily salted fish Some very aged cheeses
<0.60	Instant noodles Sugar and chocolate confectionery products Biscuits and crackers Savoury snacks, e.g. potato crisps Dehydrated vegetables Cornflakes

Table 2.7 Water activity (a_w) limits for microbial growth[a].

Organism	Minimum a_w
Campylobacter jejuni	>0.987
Vibrio cholerae	0.984
Vibrio parahaemolyticus	0.981
Staphylococcus aureus for growth and enterotoxin production	0.98
Clostridium perfringens	0.97
Eschericha coli (pathogenic)	0.95
Salmonella spp.	0.94
Bacillus cereus	0.93
Listeria monocytogenes	0.92
Most bacteria	0.91
Most yeasts	0.85
Aspergillus flavus for growth (for aflatoxin production)	0.80 (0.82)
Most moulds	0.80
Halophilic bacteria	0.75
Xerophilic bacteria	0.65
Osmophilic yeasts	0.60

[a] Source: ICMSF (1996).

The classic method for determining a_w or ERH is the Landrock and Proctor method (Cakebread, 1974). The method involves the use of a series of closed containers such as laboratory humidity ovens or dessicators. Samples of the food are exposed in these containers, each having an atmosphere of known controlled humidity created with a relative humidity control solution. (ERH values of the more commonly used saturated solutions are given in Table 2.8 (HMSO, 1970).) The changes in the sample weight are noted; the gain or loss in weight after the same time at each humidity is determined and plotted against the humidity. A curve of weight change versus relative humidity is drawn and the point at which it crosses the zero line, i.e. the point at which there is no change in weight, is the ERH. Nowadays, water activities of food can be easily and rapidly measured by instruments such as the Novasina meter manufactured by a Swiss company, Novatron Ltd. They may be calculated using, for example, the sucrose equivalence method (Cauvain & Young, 2000) or estimated using the ERH CALC™ software (see also Section 1.11).

pH value and acidity (total acidity and the type of acid)

The pH value of a food product varies according to its composition and formulation, and it needs to be controlled where acidity has a major

Table 2.8 Variation of relative humidity of air over saturated salt solutions with temperature (HMSO, 1970).

Salt	Relative humidity (%) at temperature (°C)							
	5	10	15	20	25	30	35	40
Potassium sulphate	98	98	97	97	97	96	96	96
Potassium nitrate	96	95	94	93	92	91	89	88
Potassium chloride	88	88	87	86	85	84	83	82
Potassium bromide	85	84	83	82	81	80	79	79
Ammonium sulphate	82	82	81	81	80	80	80	79
Sodium chloride	76	76	76	76	75	75	75	75
Sodium nitrite	–	–	–	66	65	63	62	62
Ammonium nitrate	74	72	69	65	62	59	55	53
Sodium dichromate	59	58	56	55	54	52	51	50
Magnesium nitrate	58	57	56	55	53	52	50	49
Potassium carbonate	–	47	44	44	43	43	43	42
Sodium iodide	43	42	40	39	38	36	35	33
Magnesium chloride	34	34	34	33	33	33	32	32
Potassium acetate	24	24	23	23	22	22	21	20
Lithium chloride	14	13	13	12	12	12	12	11
Potassium hydroxide	14	13	10	9	8	7	6	6

influence on the shelf life and safety of the product. The pH of a system is related to the concentration of hydrogen ions which, in the case of food, come from 'acid' ingredients that dissociate in water, releasing them in the process. It is a well established fact that microorganisms can only grow and multiply within certain pH ranges. Some typical food pH values are given in Table 2.9 and optimum pH ranges for growth of a number of organisms are given in Table 2.10. Again, these figures must not be taken as absolute values as real food systems are often very complex, consisting not just of an aqueous phase, but also proteins, oils and fats, and many other components. The complexity of foods has made both the accurate measurement and prediction of pH of food products difficult (Wilson & Hibberd, 2000). In addition, the pH of a food product may vary during its shelf life, as a result of changes taking place in the food, e.g. mould growth. Such a change in pH, i.e. a drop in pH, is indeed one of the preservation factors (or hurdles) in some fermented sausages and mould-ripened cheeses.

Certain acid ingredients (Table 2.11), both organic and inorganic, have specific antimicrobial effects of their own. In the case of the organic acids, their preserving effect is attributed to their undisso-

Table 2.9 Typical pH values of some foods (ICMSF, 1988).

pH range	Food	pH
Low acid (pH 7.0–5.5)	Milk	6.3–6.5
	Cheddar cheese	5.9
	Roquefort cheese	5.5–5.9
	Bacon	5.6–6.6
	Red meat	5.4–6.2
	Ham	5.9–6.1
	Canned vegetables	5.4–6.4
	Poultry	5.6–6.4
	Fish	6.6–6.8
	Crustaceans	6.8–7.0
	Butter	6.1–6.4
	Potatoes	5.6–6.2
	Rice	6.0–6.7
	Bread	5.3–5.8
Medium acid (pH 5.5–4.5)	Fermented vegetables	3.9–5.1
	Cottage cheese	4.5
	Bananas	4.5–5.2
	Green beans	4.6–5.5
Acid (pH 4.5–3.7)	Mayonnaise	3.0–4.1
	Tomatoes	4.0
High acid (pH <3.7)	Canned pickles and fruit juice	3.5–3.9
	Sauerkraut	3.1–3.3
	Citrus fruits	3.0–3.5
	Apples	2.9–3.3

Table 2.10 pH limits for microbial growth (ICMSF, 1996).

Organism	Optimum pH
Campylobacter jejuni	6.5–7.5
Vibrio cholerae	7.6
Vibrio parahaemolyticus	7.8–8.6
Staphylococcus aureus for growth (for enterotoxin production)	6–7 (7–8)
Clostridium perfringens	7.2
Eschericha coli (pathogenic)	6–7
Salmonella spp.	7–7.5
Bacillus cereus	6–7
Listeria monocytogenes	7.0
Aspergillus flavus for growth	5–8

Table 2.11 Common organic and inorganic antimicrobial acids (Gould, 1996).

Acids	Examples of foods in which used
Weak organic acid and ester preservatives	
Propionate	Bread, cakes, cheeses, grain
Sorbate	Cheeses, syrups, cakes, dressings
Benzoate	Pickles, soft drinks, dressings
Benzoate esters (parabens)	Marinated fish products
Organic acid acidulants	
Lactic, citric, malic, acetic, etc.	Low pH sauces, mayonnaise, dressings, salads, drinks, yoghurts, fruit juices and concentrates
Inorganic acid preservatives	
Sulphite	Fruit pieces, dried fruit, wine, meat sausages
Nitrite	Cured meat products
Mineral acid acidulants	
Phosphoric, hydrochloric	Soft drinks

ciated forms and they generally increase in their effectiveness in the order acetic, proprionic, sorbic and benzoic. The proportion of the acid that is in the undissociated form is determined by its dissociation constant (p*K*) and the pH of the food. Since the p*K* values of these common weak organic acids range from 4.2 (benzoic) to 4.87 (proprionic), at pH values much above these, their preservative activity is greatly reduced (Gould, 1996).

As pointed out earlier, pH also affects many chemical and biochemical changes in food, such as enzymic and non-enzymic browning, degradation of aspartame and the shade of some colours (see Sections 2.1.3 and 2.1.4). These, in turn, can have an impact on shelf life.

Redox potential (E_h)

This is generally defined as the ease with which a substrate loses or gains electrons. Oxidation involves the removal of electrons and as an element or compound loses electrons, it is oxidized. Oxidation is also achieved when an element or compound reacts with oxygen. The availability of oxygen therefore affects the oxidation–reduction (redox) potential (E_h) of a system or environment. This potential, in turn, is crucial to the survival and growth of microorganisms as well as to the many chemical and biochemical reactions in foods that require oxygen (see Sections 2.1.3 and 2.1.4). Aerobes are species capable of growth at full oxygen tensions (e.g. *Bacillus* spp.). Microaerophiles require oxygen but at levels lower than atmospheric (e.g. lactobacilli and *Campylobacter* spp.). Facultative aerobes (e.g. *Escherichia coli*) can grow under either aerobic or anaerobic conditions. *Clostridium perfringens* is an obligate anaerobe that cannot survive in the presence of oxygen.

Redox potential is one of the most important preservation factors (hurdles) employed in the manufacture of ready-to-eat and ambient stable meat products in Germany. The careful adjustment and proper control of a number of critical parameters including E_h are essential for assuring the shelf lives of shelf stable products (SSPs) like bologna-type sausages (F-SSP) and brühdauerwurst (a_w-SSP) (Leistner, 2000).

2.2.2 Extrinsic factors

Processing

Processing covers a wide range of operations to which food may be subjected. It can exert a considerable effect on the microflora, physical, chemical, biochemical, nutritional and sensory properties of a food product, and hence its shelf life. Some of the common processing operations are listed below.

Raw material preparation

Examples are sorting, washing and peeling.

Size reduction

Examples are slicing, chopping and grinding.

Separation

Examples are sieving, filtration and centrifugation

Other operations that can be carried out under ambient conditions

Examples are mixing, forming and fermentation.

Operations that use steam or hot water

Examples are blanching, pasteurization, sterilization, evaporation, cooking and extrusion-cooking.

Operations that use hot air or hot oil

Examples are drying, baking, roasting and frying.

Operations that remove heat from food

Examples are chilling and freezing.

Other ancillary and post-processing operations

Examples include enrobing and coating, filling, sealing, packaging and metal detecting.

Although many of these operations, e.g. pasteurization and chilling, do play a shelf life influencing role, food processing operations should not be considered in isolation. In addition to the processing they have been given, most food products rely on pack integrity to achieve their assigned shelf lives. The severity of processing, too, can be determined by other factors. For example, it is unnecessary to heat foods that are more acid than pH 4.5 to the same extent as 'low acid foods' that have a pH higher than 4.5. In the absence of other preservation factors, the latter will require a 'botulinum cook' as a minimum heat treatment if a shelf life in excess of 6 months at ambient temperatures is to be achieved. In view of the influence processing can have on shelf life, the critical processing stage in food manufacture must be identified and its associated conditions (e.g. the time and temperature combination) established using HACCP. Furthermore, the corresponding critical control point(s) will need to be managed effectively during routine production.

Hygiene

Good hygiene, an integral part of GMP, is fundamental to the manufacture of safe and wholesome food products. In a sense, the requirement of food hygiene distinguishes food from non-food manufacture. Poor hygiene leads to contamination which may be physical, chemical or microbiological in nature, and which can have a major impact on the safety and stability of foods. Research has clearly demonstrated that poor slicing hygiene caused the spoilage (and reduced shelf life) of some chilled vacuum-packed cured meats (Holley, 1997) and that the in-house flora had a definite impact on the microbiological quality and shelf life of cold-smoked salmon (Hansen *et al.*, 1998). Past experience has indicated that the mould-free shelf lives of some cakes can be extended by 10–15% simply as a result of better hygiene, improved housekeeping, and more effective pest control in the bakery (Jones, 2000b). However, hygiene as a shelf life influencing factor is not usually included in storage trials and is not one of the variables available within current predictive microbiological models.

Packaging materials and systems

Packaging protects and preserves. It protects food against physical damage and prevents contamination during transport, storage and distribution. Suitable packaging materials also offer a barrier against light, gaseous exchange and/or moisture vapour transfer, protecting the food from many of the deteriorative changes that can be shelf life limiting (see Sections 2.1.1, 2.1.2 and 2.1.3). Packaging, too, is often an integral part of a food preservation system. Examples are metal cans for canned foods and laminate cartons for aseptically processed drinks. In recent years, many food developments have been led by innovations in packaging, the adoption of which often results in an extension of shelf life (see also Section 1.13). Table 2.12 gives some examples of foods packed in MAP, their approximate shelf lives and the recommended gas mixes.

Storage, distribution and retail display

Conditions such as temperature, relative humidity, light exposure and degree of physical handling during storage, distribution and retail

Table 2.12 Modified atmosphere packaging (MAP) of some foods and their achievable shelf lives (Anon., 1995)[a].

Food category	Typical gas mix for retail packs			Achievable shelf life (days)	
	O_2(%)	CO_2(%)	N_2(%)	In air	In MAP
Raw red meat	70	30		2–4	5–8
Raw poultry		30	70	4–7	10–21
Raw, low fat white fish and seafood	30	40	30	2–3	4–6
Cooked, cured and processed meat products		30	70	7–21	21–49
Cooked, cured and processed fish and seafood products		30	70	5–10	7–21
Ready meals and cook-chill products		30	70	2–5	5–10
Fresh pasta products		50	50	7–14	21–28
Fresh prepared fruit and vegetable products	5	5	90	2–7	5–35
Dairy products		100		7–28	14–84
Bakery products (ambient)		50	50	4–14	28–84
Dried food products (ambient)			100	4–8 months	1–2 years

[a] Unless otherwise stated, recommended storage temperature is 0–3°C.

display of food products can have a marked effect on their shelf lives. Sometimes because they are not known or cannot be easily determined, the temptation is not to take them into account in determining shelf life. It is possible to predict bacterial growth in chilled products using data obtained by logging temperatures through a chill distribution system. Figure 2.1 shows results from such a set of data (Baird-Parker & Kilsby, 1987). Research has been conducted to examine the influence of the thermal history (i.e. during storage and distribution) on the shelf life of carbonated beverages in plastic containers. It was found that under conditions comparable to those occurring during real-life distribution, the shelf life estimated differed significantly from that predicted by neglecting the temperature rise due to outdoor storage and sunlight exposure (Del Nobile *et al.*, 1998). Emulsified

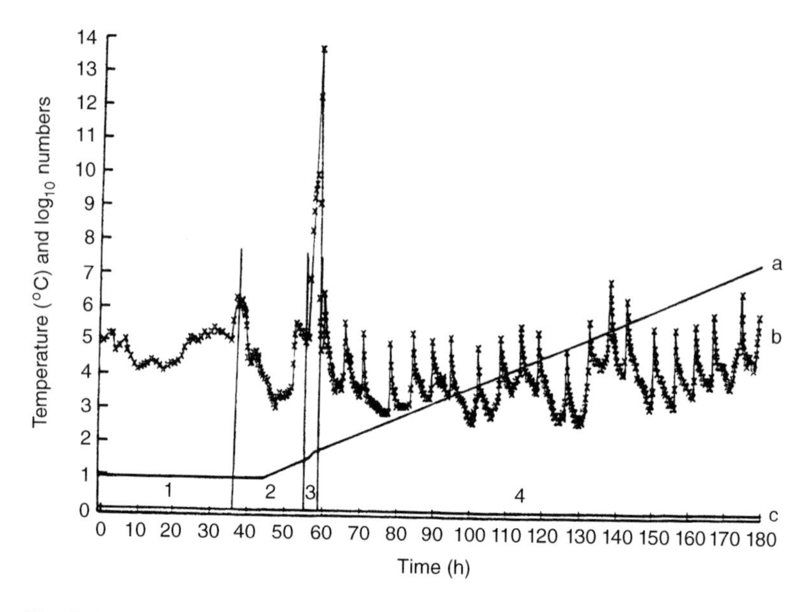

Fig. 2.1 Temperature profile during distribution and bacterial growth: curve a, prediction of growth of Gram-negative meat spoilage organisms; curve b, temperature variations; curve c, prediction of growth of *Staphylococcus aureus*.
Key to areas on graph: 1, producers despatch refrigeration; 2, in distribution to retail store via depot; 3, in-store handling before chill; 4, in store refrigerated cabinet. (Reproduced from Baird-Parker & Kilsby (1987) with permission from *Journal of Applied Bacteriology*.)

sauces such as mayonnaise have been known to suffer from emulsion breakdown during distribution as a result of excessive vibration exacerbated by high ambient temperatures. Thus, conditions during storage and distribution should not be ignored. For certain types of food products, e.g. chilled foods in the UK, the relevant regulations allow a maximum of 8°C during distribution and retail display (HMSO, 1995b). In this case, the conditions must be simulated and included during storage trials. One development that is expected to make a contribution to the monitoring of the temperature history of food products during storage, distribution and retail display involves the use of time–temperature indicators (TTIs). These are sensors that can provide information related to the temperature history of the products to which they are attached. TTIs may be classified as follows (Singh, 2000):

- Critical temperature indicators, e.g. MonitorMark™ Thaw Indicator manufactured by 3M™ Packaging Division.
- Partial history indicators, e.g. MonitorMark™ Threshold Indicator manufactured by 3M™ Packaging Division.
- Full history indicators, e.g. LifeLines™ Fresh-Scan Indicator manufactured by LifeLines Technologies.

Detailed specifications for TTIs covering areas such as technical performance (accuracy and reproducibility) and reference testing have been published (BSI, 1999; George & Shaw, 1992).

2.2.3 Interaction between intrinsic and extrinsic factors

Both intrinsic (e.g. product composition) and extrinsic (e.g. processing) factors can assert their influence on shelf life independently or interactively. There are many examples of interaction between these factors. For instance, chocolate products have to be reformulated for export to tropical countries because cocoa butter melts fully and completely just below body temperature, i.e. at 36°C. Canned foods destined for a hot climate will have to receive a more severe thermal process to prevent spoilage due to thermophilic bacteria. The use of curing salts that gives chilled bacon its shelf life will increase the risk of lipid oxidation due to concentration effect if the bacon is frozen. In Germany, the popular raw mini-salami is a fermented product that relies on a number of hurdles to give it ambient stability. Any pathogenic organisms that may survive the manufacturing process quickly die under ambient storage conditions. It is believed that if the product is stored under refrigerated conditions, the pathogens will live much longer, making the product unsafe (Leistner, 2000). In determining shelf life, it is important therefore to evaluate the effects of all the relevant factors individually and in combination.

2.2.4 Consumer handling and use

As far as the shelf life of foods is concerned, potentially this can be an important shelf-life determining factor and yet it is highly variable and one over which the manufacturer has little control. For a given product, there are in theory an endless number of shopping and carry-home patterns, the duration and conditions of which are also difficult to predict. Once at home, there are again many possible combinations and permutations of storage, handling and use. A certain amount of information concerning the handling of chilled foods is available (Evans *et al.*, 1991; Evans, 1998). However, given finite resources, it is difficult to see how consumer handling and use can be simulated adequately in storage trials. It follows, too, that it is impracticable to mimic abused conditions in shelf life studies as they can be unpredictable. In practice, the inclusion of a 'worst case' storage regime (see also Section 1.14.2) in every storage trial and a generous margin of safety in the final shelf life assigned are precautions that the manufacturer can take. In the long run, education of consumers through clear instructions on the label, information leaflets, etc., in how the product should be handled, stored and used will become an increasingly important tool in the management of shelf life. Indeed, the provision of 'any special storage conditions or conditions of use' is already a general labelling requirement (see Section 1.5). The following list gives some commonly seen label advice pertinent to storage and/or use at home:

- 'Store in a cool, dry place; once opened keep refrigerated and use within 4 weeks' (e.g. tomato ketchup).
- 'Store in a cool dry place, out of direct sunlight' (e.g. ambient rich baked goods with significant fat content).
- 'Store in the refrigerator' (e.g. margarine and reduced-fat spread).
- 'Keep refrigerated 0°C to +5°C; suitable for freezing: freeze on day of purchase, use within one month, defrost thoroughly before use' (e.g. ready-to-cook chilled foods).
- 'Cook from frozen', 'once defrosted, do not refreeze' (e.g. some frozen foods).

2.2.5 Commercial considerations

The final shelf life that is specified is often affected by commercial considerations, some of which are:

- The level of stocks required to sustain a given distribution and retail chain without the danger of running out of products on display.
- The need to meet seasonal demands.
- Occasional requirements to stockpile in support of special promotions.

Major retailers nowadays often require a minimum residual shelf life for branded products at the point of delivery as part of the agreement with their suppliers, whereas the designated shelf life of own label products is often much less than the branded equivalent because own label products often enjoy shorter turnover times resulting from a dedicated and efficient distribution network and competitive pricing policy.

2.3 Summary

The following are a number of the key points:

- Many of the mechanisms of food deterioration and spoilage are well known.
- The biochemistry, chemistry and microbiology of food provide much of the root on which our current understanding of food spoilage and deterioration is based.
- There are numerous factors, intrinsic and extrinsic, that influence how food spoils. These factors can act independently or interactively.
- Consumer handling and use, over which manufacturers often have little control, can be an important shelf life influencing factor.
- The shelf life finally assigned to a food product may be decided dependent upon commercial circumstances, resulting in a shelf life that is significantly shorter than is otherwise possible based on technological considerations alone.

Section 3

Determining Shelf Life in Practice

This section uses three case studies to illustrate how shelf life may be determined in practice. The examples are fictional and there may well be alternative or better procedures and methodology. Nonetheless, they are intended to show what is possible and are provided without any liability whatsoever in their application and use.

3.1 Short shelf life products

3.1.1 The product

This is a chilled product. It consists of sliced fresh potatoes in a white sauce, topped with grated cheese. The product is fully cooked and requires heating according to predetermined instructions before serving. The declared pack weight is 310 g.

3.1.2 The process

Fresh potatoes of a specified variety are abrasion-peeled. Peeled potatoes are inspected and hand-trimmed if necessary. They are then machine-sliced (e.g. Urschel slicer) to a specified thickness (e.g. 2 mm). Potato slices may be held in 2.5% salt (NaCl) solution in a chill store before use. The white sauce is made by mixing fresh full-cream milk, modified corn starch, ground white pepper and salt, using a suitable homogenizer. Potato slices are drained immediately before use, and hand-filled into aluminium trays into which a specified amount of white sauce is machine-filled. Into each of the trays, a weighed amount of grated cheese is sprinkled. Products are baked in a gas oven to reach a product centre temperature of 95°C. Cooked products are chilled within 30 minutes of leaving the oven to below 5°C in an air blast chiller within a further period of 90 minutes. Chilled products are covered and closed with a preprinted cardboard lid coded with a 'use by' date. The product is stored, distributed and retail-displayed under refrigerated conditions to comply with the relevant UK regulations, i.e. maximum of 8°C. An outline of the process is given in Fig. 3.1.

3.1.3 Food safety

In order to assure product safety, a HACCP study is carried out based on the principles published by the Codex Commission on Food Hygiene (Codex, 1997). The aim is to identify the relevant chemical, microbiological and physical hazards and their associated critical control points (CCPs). The study is repeated before product launch to confirm the CCPs and the monitoring and control procedures.

It is further assumed that the manufacturing operations of the company in question have been designed, carried out and managed according to the GMP principles detailed in the IFST guide (Blanchfield, 1998).

3.1.4 Mechanism(s) of shelf life deterioration

This is a moist prepared food. It is cooked but not sterile. If contaminated after cooking, it can support microbial growth that will either cause food poisoning or spoilage. The product is basically preserved by hygienic post-process handling and temperature control from factory through distribution to home use. However, the chill chain alone cannot be relied upon to prevent microbial growth so that the product will have a limited microbiological shelf life even at refrigerated temperature. It is therefore a high-risk product. The eating quality of the product deteriorates on storage. Based on published information, in-house technical expertise and past

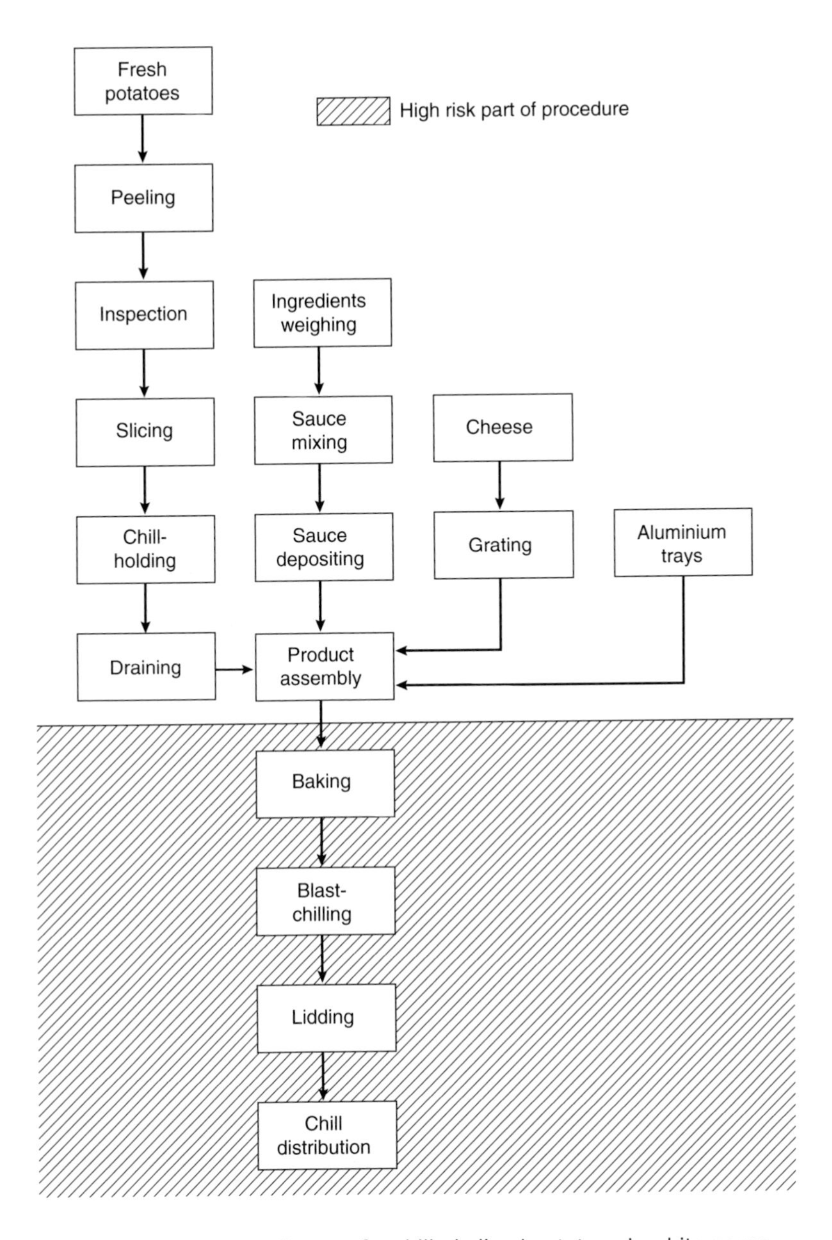

Fig. 3.1 Process flow diagram for chilled sliced potatoes in white sauce.

experience, the probable mechanisms of shelf life deterioration are identified as:

- Microbial growth.
- Deterioration in flavour and texture, likely to be caused by physicochemical changes.

3.1.5 *Shelf life determination – storage trials*

The objective of the investigation is to determine an acceptable shelf life of the product.

Storage conditions

- Control: deep frozen at −18°C.
- Average: 0–5°C.
- Worst: 8°C.

Storage duration

- 10 days (to be kept refrigerated between 0°C and +5°C) from day of cooking, although only 8 days are required.

Sampling plan: schedule, samples and tests

The plan is to evaluate the product daily, a total of 11 sampling occasions. The tests used to assess storage behaviour of the products are:

- Microbiological examination. The tests are for aerobic colony count (ACC) at 30°C for 48 hours and for Enterobacteriaceae (IFST, 1999). These are to be carried out on samples taken on days 0, 4, 8 and 10.
- Sensory evaluation. Four employees of the company, who are familiar with the product, form the taste panel. They use a seven-point hedonic scale (1 = completely unacceptable, 4 = moderately acceptable, 7 = as acceptable as control) to evaluate appearance, flavour, texture and overall acceptability

of the product. Alternatively, an appropriate difference test can be used to evaluate 'average' and 'worst case' samples against the 'control'. Once a significant difference has been detected, an appropriate hedonic test can be used to determine the sensory end-point. Ideally, a sensory specification that has been agreed with the customer or developed based on market research should be used.

The minimum number of packs required is calculated below:

- Microbiological examination. Because there are four sampling occasions and three storage temperatures, ten packs are required as day 0 does not have three temperatures.
- One pack is needed for the taste panel and so 31 (i.e. $(3 \times 11) - 2$) packs are required over the storage period.
- A total of at least 41 packs is needed, but it is prudent to lay down more samples to give a safety margin (e.g. by doubling the number of packs kept for the taste panel).

It is good practice to analyse the data as they become available and, if appropriate, plot the results on a graph to highlight any trends or anomalies. Also, at some stage, a travel test should be conducted with products packed in secondary packaging designed for routine production.

3.1.6 Predicting shelf life

If the pH and salt content (% w/v) of the product are known, the microbiological shelf life of the product may be estimated using, for instance, the Forecast service offered by CCFRA. Given the Enterobacteriaceae count is known on day 0 (e.g. 100 cfu/g), the time (days) taken to reach a level of 10^4 cfu/g at the point of sale (Gilbert *et al.*, 2000) at a given temperature (e.g. 8°C) can be modelled. Because this is a cooked as well as a multicomponent product, it is strongly advised that help should be sought from an experienced food microbiologist regarding the prediction of its microbiological shelf life and interpretation of the results.

3.1.7 Assurance of assigned shelf life – the HACCP approach

In any analysis based on the HACCP principles, the following are essential:

- Identification of major hazards: a hazard, in this case, is the potential to cause a reduction in shelf life.
- Determination of the critical control points (CCPs), establishment of their critical limits and of their monitoring procedures.

In this example, the major hazards are:

- Microbial growth.
- Deterioration in flavour and texture.

Examples of CCPs are:

- The supply of potatoes. Potatoes should be purchased to a specified variety and an appropriate specification. Test cooking should be carried out using samples from every delivery. The maximum storage life before processing under defined conditions should also be specified. An appropriate procedure and recording system should be put in place to manage this CCP.
- The thickness of potato slices. This needs to be controlled and monitored because it affects the quality and shelf life of the product. A suitable measuring template can be made to gauge the thickness of, say, ten slices.
- The cooking process. The time and temperature combination of this process needs to be controlled and monitored. Additionally, product centre temperature of cooked samples should be measured to ensure it is at 95°C.
- The chilling process. This process needs to be controlled and monitored to ensure that the standing time before chilling does not exceed 30 minutes and that the products are brought down to 5°C in 90 minutes. A suitable record form can be designed for use at this CCP.

3.2 Medium shelf life products

3.2.1 The product

The product is ready salted flat-cut potato crisps packaged in bags produced from fully printed metallized oriented polypropylene (OPP) film. The declared pack weight is 50 g.

3.2.2 The process

Fresh potatoes of a specified variety are abrasion-peeled. Peeled potatoes are inspected and hand-trimmed if necessary. They are then machine-sliced (e.g. Urschel slicer) to a specified thickness (e.g. 1.5 mm). Potato slices are washed to remove surface starch and allowed to drain-dry. Slices are cooked in a batch deep fat fryer at 180°C for 2.5 minutes. Cooked crisps are drained of surplus oil, salted and packaged in OPP film already coded with a 'best before' date. An outline of the process is given in Fig. 3.2.

3.2.3 Food safety

In order to ensure product safety, a HACCP study is carried out based on the principles published by the Codex Commission on Food Hygiene (1997). The aim is to identify the relevant chemical, microbiological and physical hazards and their associated critical control points (CCPs). The study is repeated before product launch to confirm the CCPs and the monitoring and control procedures.

It is assumed that the manufacturing operations of the company in question have been designed, carried out and managed according to the GMP principles detailed in the IFST guide (Blanchfield, 1998).

3.2.4 Mechanism(s) of shelf life deterioration

Typically, the product has a moisture content of less than 1% and an oil content of 30% after frying. Since the a_w of the product is less than 0.6 (see Table 2.6), spoilage due to microbiological changes is

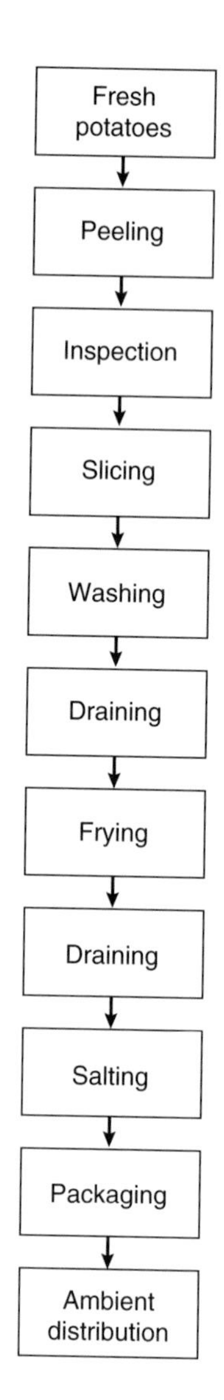

Fig. 3.2 Process flow diagram for ready salted potato crisps.

unlikely. Based on published information, in-house technical expertise and past experience, the probable mechanisms of shelf life deterioration are identified as:

- Oxidative rancidity.
- Moisture uptake.

3.2.5 Shelf life determination – storage trials

The objective of the investigation is to determine an acceptable shelf life of the product.

Storage conditions

- Control: deep frozen at −18°C.
- Average: 20°C, 55%, i.e. ambient conditions.
- Worst: 25°C, 75%, i.e. temperate conditions.

Storage duration

- 8 weeks (to be stored in a cool dry place); the customer only wants 6 weeks.

Sampling plan: schedule, samples and tests

The plan is to evaluate the product at weekly intervals so that samples will be taken off for various tests at weeks 0, 1, 2, 3, 4, 5, 6, 7 and 8.

The tests used to assess storage behaviour of the products are:

- Determination of peroxide value (PV) of the extracted oil – PV is accepted as an indication of the extent of oxidative rancidity.
- Weighing of the packs – the increase in weight is taken to represent moisture uptake of the product.
- Sensory evaluation. Six employees of the company who are familiar with the product form the taste panel. They use a seven-point hedonic scale (1 = completely unacceptable, 4 = moderately acceptable, 7 = as acceptable as control) to evaluate flavour, texture and overall acceptability of the product. Alternatively, an appropriate difference test can be used to evaluate 'average' and 'worst case' samples against the 'control'. Once a significant difference has been detected, an appropriate hedonic test can be used to determine the sensory end-point. Ideally, a sensory specification that has been agreed with the customer or developed based on market research should be used.

The minimum number of packs required is calculated below:

- Weighing and PV determination are carried out on the same duplicate samples; because there are nine sampling occasions and three storage temperatures, 50 packs are required (week 0 does not have three storage temperatures);
- One pack is needed for the taste panel and so 25 packs are required over the storage period.
- A total of at least 75 packs is needed, but it is prudent to lay down more samples to give a safety margin (e.g. by doubling the number of packs kept for the taste panel).

All samples are weighed and labelled before being put into storage. It is good practice to analyse the data as they become available and, if appropriate, plot the results on a graph to highlight any trends or anomalies. Also, at some stage, a travel test should be conducted with products packed in secondary packaging designed for routine production.

3.2.6 *Assurance of assigned shelf life – the HACCP approach*

In any analysis based on the HACCP principles, the following are essential:

- Identification of major hazards; a hazard, in this case, is the potential to cause a reduction in shelf life.
- Determination of the critical control points (CCPs), establishment of their critical limits and of their monitoring procedures.

In this example, the major hazards are:

- Oxidation of the oil that will lead to rancidity development.
- Moisture pick-up that will result in loss of crispness.

Examples of CCPs are:

- The supply of frying oil. The oil is purchased to a quality specification and every delivery is accompanied with a certificate of analysis.

- The frying process. There are at least two aspects to this process. The time and temperature combination (180°C for 2.5 minutes) of the process needs to be controlled. As it is known that decrease in quality of the frying oil on continuing use contributes to decrease in sensory acceptability of potato crisps, the quality of the frying oil needs to be controlled and monitored. This can be done by determining the free fatty acid content of the oil in the fryer, or measuring its viscosity daily or at appropriate intervals.
- Handling after frying. After frying, the crisps have a very low moisture content. The duration between end of frying and packaging therefore needs to be controlled and monitored to prevent unnecessary moisture pick-up by the crisps. A maximum duration needs to be specified. Then it can be monitored, for instance, by the use of a process sheet that records the time when frying of each batch ends and the time when the packaging of it begins.
- The pack seal integrity. All the good work that has gone into producing crisps of specified quality will be ruined if the pack seal is defective. Depending on the throughput, finished packs need to be sampled regularly and evaluated for seal integrity.

3.3. Long shelf life products

3.3.1 The product

The product is a frozen dairy cream sponge sandwich. It is flow-wrapped in clear polypropylene film and packed in a printed cardboard carton. The product is a golden sponge with dairy cream and raspberry jam sandwiched in it. The net product weight is 260 g.

3.3.2 The process

Weighed ingredients (wheat flour, sugars, pasteurized liquid egg, margarine, whole milk powder, baking powder, water, and colours, etc.) are high-speed mixed and aerated to produce the sponge batter. Batter is deposited volumetrically into pregreased baking tins. Filled tins are baked in a gas-fired oven. Baked sponge bases are cooled, removed from the tins and cut horizontally. Onto the bottom halves are deposited jam and whipped dairy cream, and the top halves are put back. The assembled sponge sandwiches are blast-frozen, flow-wrapped, cartoned and coded with a 'best before' date. An outline of the process is given in Fig. 3.3.

3.3.3 Food safety

To ensure product safety, a HACCP study is carried out based on the principles published by the Codex Commission on Food Hygiene (1997). The aim is to identify the relevant chemical, microbiological and physical hazards, and their associated critical control points (CCPs). The study is repeated before product launch to confirm the CCPs and the monitoring and control procedures.

It is assumed that the manufacturing operations of the company in question have been designed, carried out and managed according to the GMP principles detailed in the IFST guide (Blanchfield, 1998).

3.3.4 Mechanism(s) of shelf life deterioration

After freezing and during frozen storage, microbial growth is either very much slowed down or totally arrested. Although freezing and subsequent frozen storage cannot be regarded as bacteria-killing processes, microbiological spoilage associated with frozen foods is uncommon as long as the product has been manufactured hygienically and according to GMP principles. In the present example, this is a particularly important consideration as the dairy cream filling is a high-risk component and can be hazardous if it is not handled correctly. Despite frozen storage, the probable mechanisms of deterioration are identified as:

- Physical changes such as moisture redistribution, migration and loss between the different components, i.e. sponge, dairy cream and jam.
- Physicochemical changes such as staling of sponge.

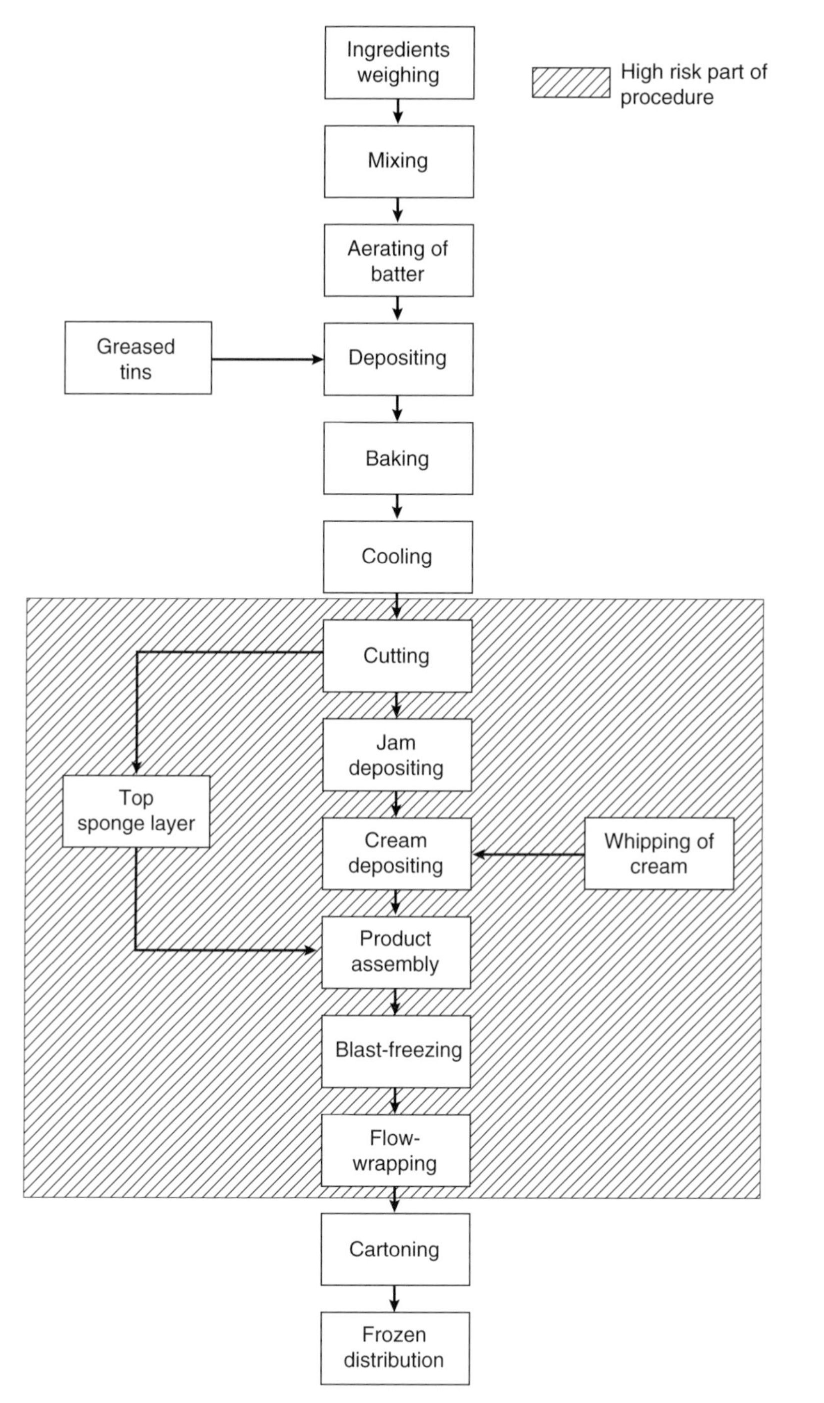

Fig. 3.3 Process flow diagram for frozen sponge sandwich with dairy cream and jam filling.

3.3.5 *Shelf life determination – storage trials*

The objective of the investigation is to determine an acceptable shelf life of the product.

Storage conditions

- Control: deep frozen at −30°C.
- Average: at −18°C (food freezer or three-star marked frozen food compartment).
- Worst case or accelerated:
  −12°C (two-star marked frozen food compartment)
  −6°C (one-star marked frozen food compartment)
  −2 to 0°C (ice making compartment in a refrigerator).

Storage duration

- 18 months (to be kept frozen at −18°C etc., not to be refrozen once defrosted), the customer only wants 12 months.

Sampling plan: schedule, samples and tests

The plan is to evaluate the product at bimonthly intervals so that samples will be taken off for various tests at months 0, 2, 4, 6, 8, 10, 12, 14, 16 and 18, for samples stored at control and average conditions. Additionally, samples will be taken for assessments after 3 days, 1 week and 1 month storage in an ice making compartment, one-star marked and two-star marked frozen food compartments, respectively.

The tests used to assess storage behaviour of the products are:

- Weighing of the unwrapped products to monitor any moisture loss; this can be done immediately before sensory evaluation.
- Sensory evaluation. Five employees of the company who are familiar with the product form the taste panel. They use a seven-point hedonic scale (1 = completely unacceptable, 4 = moderately acceptable, 7 = as acceptable as control) to evaluate appearance, flavour, texture, overall acceptability and

freeze–thaw stability of the product. Alternatively, an appropriate difference test can be used to evaluate 'average' and 'worst case' samples against the 'control'. Once a significant difference has been detected, an appropriate hedonic test can be used to determine the sensory end-point. Ideally, a sensory specification that has been agreed with the customer or developed based on market research should be used.

The minimum number of packs required is calculated below:

- One pack from each set of the control and average conditions is needed for the taste panel; so 19 packs are required over the 18 month storage period.
- Three packs from each set of the worst case or accelerated conditions are needed for the panel; so nine packs are required.
- A total of at least 28 packs are needed, but it is prudent to lay down more samples to give a safety margin (e.g. by doubling the total number of packs kept).

Samples for the control and average conditions are weighed and identified before they are boxed and put into storage. It is good practice to analyse the data as they become available and, if appropriate, plot the results on a graph to highlight any trends or anomalies. Also, at some stage, a travel test should be conducted with products packed in secondary packaging designed for routine production.

3.3.6 Assurance of assigned shelf life – the HACCP approach

In any analysis based on the HACCP principles, the following are essential:

- Identification of major hazards; a hazard, in this case, is the potential to cause a reduction in shelf life.
- Determination of the critical control points (CCPs), establishment of their critical limits and of their monitoring procedures.

In this example, the major hazards are:

- Staling of the sponge, perceived as increased hardness and dryness.
- Physical deterioration such as moisture loss, colour migration from jam into sponge and dairy cream.

Examples of CCPs are:

- The pack seal integrity. All the good work that has gone into producing the product of specified quality will be ruined, particularly over a very long period of storage, if the pack seal is defective. Depending on the throughput, finished packs need to be sampled regularly and evaluated for seal integrity.
- The cold chain. Because the stability of the product critically depends on the integrity of the cold chain, the temperature profile of the product throughout the chain needs to be monitored regularly to identify and prevent any temperature abuse. In the UK, in deciding the monitoring programme, due regard should be given to the *Quick-frozen Foodstuffs (Amendment) Regulations* (HMSO, 1994) implementing the provisions of Commission Directives 92/1/EEC and 92/2/EEC (EEC, 1992a, b), respectively, concerning:
    1. the monitoring of temperatures of quick-frozen foodstuffs during transport, warehousing and storage; and
    2. the sampling procedure and methods of temperature measurement for quick-frozen foodstuffs.

3.4 Summary

The following are a number of the key points:

- The classification of food products into short, medium and long shelf-life products is arbitrary.
- In the UK, from a legal and labelling point of view, there are basically two categories of food products: one for which a 'use by' date is appropriate and the other, a 'best before' date.

- The most common and direct way of determining shelf life is to conduct carefully planned and executed storage trials under defined conditions.
- Shelf life determination should consider all stages in the manufacture of a food product.
- Information and data about raw materials used, product composition and formulation, product assembly, processing techniques, packaging used, environmental hygiene, storage and distribution procedures, and consumer handling and use are all important considerations.
- Only when all these areas have been considered will a meaningful and reliable determination be possible.
- The principles of HACCP are equally applicable to the identification of hazards that reduce shelf life and the management of the CCPs associated with them.

Epilogue

The subject of shelf life is complex but fascinating. The determination of shelf life should be a multidisciplinary endeavour. Assurance of food safety and specified quality are the two main aspects of an acceptable shelf life. Issues relating to safety, however, must always take precedence over those relating to quality. Storing product samples under defined conditions remains the most common approach to the shelf life determination of foods. Understanding the pertinent mechanism of food deterioration and the factors that affect it is fundamental to the setting as well as the extension of shelf life. Modern tools, principally in the form of predictive models for microbiological safety and stability, are now available. Development of more accurate, sophisticated and sector-specific models is ongoing. In the long run, it is hoped that increased use of predictive models will contribute to greater confidence in the microbiological shelf life of foods, stimulate wider application of predictive microbiology and reduce somewhat the strain on the conventional resource-intensive approach to determining shelf life.

A book of this size cannot hope to cover in detail the length, breadth and depth of shelf life. If it has made you think more about the subject, want to learn more about shelf life determination or to improve your existing procedures and methods, then it has been a success.

The information base on shelf life is vast, covering many journals, conference proceedings, books and book chapters. The following are a few titles which readers may find useful.

Shelf-life Dating of Foods (Labuza, 1982)
Published in 1982, this is arguably the first significant volume on the subject of shelf life. It contains 22 chapters, covering 'Basic food

preservation and degradation modes', 'Scientific investigation of shelf-life', 'Shelf-life of breakfast cereals', 'Shelf-life of bakery products', 'Shelf-life of dehydrated products', 'Shelf-life of frozen convenience products' and so on.

Evaluation of Shelf Life for Chilled Foods (CCFRA, 1990)
This is Technical Manual No. 28 published by Campden & Chorleywood Food Research Association Group, UK. The document, first produced in 1990 by a Shelf Life Working Party of the Chilled Foods Panel of the Association, was part-revised in 1997. It is intended to be used as an outline structure for the evaluation of shelf life of chilled foods including ingredients and products for retail sale. Throughout the document extensive reference is made to the HACCP approach to process assessment and control.

Shelf Life of Foods – Guidelines for its Determination and Prediction (IFST, 1993)
This is a publication of the IFST (UK). It was written on behalf of the Institute by an *ad hoc* working group consisting of members of the IFST. The aim of this publication is to provide concise advice to food business managers on the principles of shelf life determination and prediction, at every point in the food chain. It also explains the factors influencing the shelf life of foods and the various mechanisms of deterioration in foods, which form the basis of the scientific principles essential to all evaluation of shelf life of foods. The booklet ends with a list of references.

Shelf Life Studies of Foods and Beverages – Chemical, Physical and Nutritional Aspects (Charalambous, 1993).
This is a substantial reference (1204 pages) consisting of 40 chapters by 89 contributors. The coverage ranges from shelf lives of food commodities (e.g. meat, fish, fruit and vegetables) to food (e.g. confectionery, bakery, extruded) and drink (e.g. tea, coffee, wines) products. It is an update of a similar title, published seven years earlier, from the same editor and publisher. Understandably, the bibliography is extensive.

Predictive Microbiology: Theory and Application (McMeekin *et al.*, 1993)
This is the first significant volume on predictive microbiology. It con-

sists of ten chapters. Topics covered include 'Basic concepts and methods', 'Modelling temperature effects', 'Modelling the combined effect of temperature, water activity, and other factors on microbial growth rate', 'Electronic temperature function integration' and so on. This is an excellent book and has become an essential reference for food microbiologists and modellers working in predictive food microbiology.

Shelf Life Evaluation of Foods (Man & Jones, 2000)
This multi-authored reference, now in its second edition, concentrates primarily on the shelf life of foods, although much of what is covered is equally applicable to drinks. The book, divided into two parts, begins with six chapters reviewing the principles of shelf life evaluation. The remaining ten chapters illustrate the practice of shelf life evaluation using a number of selected product groups: chilled yoghurt and other dairy desserts, fresh and lightly preserved seafood, ambient packaged cakes, potato crisps and savoury snacks, chocolate confectionery, ready-to-eat breakfast cereals, processed foods in containers other than cans, ambient-stable sauces and pickles, frozen foods and minimally processed, ready-to-eat, and ambient-stable meat products.

The Stability and Shelf-life of Food (Kilcast & Subramaniam, 2000)
This is a multi-authored reference edited by two scientists at the UK Leatherhead Food Research Association. After an introduction, the book divides into two parts. Part 1 consists of five chapters, looking at ways of analysing and predicting shelf life. Part 2 consists of seven chapters of case studies, covering 'Predicting packaging characteristics to improve shelf-life', '*Sous vide* products', 'Milk and milk products', 'Confectionery products', 'Fruits and vegetables', 'Fats and oils' and 'Sauces and dressings'.

Appendices

Appendix A
The Arrhenius Model

With very few exceptions, the rate of a chemical reaction increases, often very sharply, with increase in temperature. The relation between the rate constant k and temperature was first proposed by Svante Arrhenius in 1889:

$$k = A\exp\left(-\frac{E_a}{RT}\right)$$

The constant A is called the frequency factor, or pre-exponential factor; E_a is the activation energy; R is the universal gas constant (0.001987 kcal mol^{-1} K^{-1} or 8.31 J mol^{-1} K^{-1}) and T is the absolute temperature in K (kelvin). Converting this relationship to logarithmic form, the following is obtained:

$$\log_{10} k = \log_{10} A = \frac{E_a}{2.303RT}$$

or

$$\ln k = \ln A - \frac{E_a}{RT}$$

In theory, a plot of ln k versus the reciprocal of absolute temperature should give a straight line, the slope of which is the activation energy divided by the gas constant (E_a/R). A graph of ln k against I/T is called an Arrhenius plot; many reactions have been found to show Arrhenius behaviour, i.e. their Arrhenius plots show a straight line. Thus, by studying a reaction and measuring k at two or three different temperatures, one could extrapolate with a straight line to a lower temperature and predict the rate at this temperature. This is the basis of ASLD at an elevated temperature. Using the experimental data generated in studies by Bell & Labuza (1994), Taoukis *et al.* (1997) demonstrated that aspartame degradation showed Arrhenius behaviour in a pH 6.67 dairy system.

In the study of shelf life, another parameter that is often used to describe the relationship between temperature and reaction rate is the Q_{10} factor. Q_{10} is defined as:

$$Q_{10} = \frac{\text{Rate at temperature } (T + 10)^{\circ}\text{C}}{\text{Rate at temperature } T^{\circ}\text{C}}$$

$$= \frac{\text{Shelf life at } T^{\circ}\text{C}}{\text{Shelf life at } (T + 10)^{\circ}\text{C}}$$

Further information about Arrhenius kinetics and ASLD is available (Labuza & Riboh, 1982; Labuza & Schmidl, 1985; Mizrahi, 2000).

Appendix B The CIMSCEE Formulae for Microbiological Safety and Stability

The CIMSCEE formulae are as follows:

$$15.75(1-\alpha)(\text{total acetic acid\%}) + 3.08(\text{salt\%}) + (\text{hexose\%}) + 0.5(\text{disaccharide\%}) + 40(4.0-\text{pH}) = \Sigma_s \quad (1)$$

For any sauce based on acetic acid, if the value of Σ_s exceeds 63, safety from microbial pathogens is assured.

$$15.75(1-\alpha)(\text{total acetic acid\%}) + 3.08(\text{salt\%}) + (\text{hexose\%}) + 0.5(\text{disaccharide\%}) = \Sigma \quad (2)$$

For any sauce based on acetic acid, if the value of this formula (Σ) exceeds 63, microbial spoilage should not occur.

$(1-\alpha)$ is the proportion of the total acetic acid that is undissociated and is related to the pH of the product and the pK of acetic acid, where

$$\text{pH} = \text{p}K + \log_{10}(\alpha/1-\alpha)$$

and

$$\text{p}K_{\text{acetic acid}} = 4.757$$

These formulae do not:

- Apply to pickles or dressed salads.
- Take into account the effects of any other antimicrobial components present, e.g. added preservatives or natural anti-

microbial agents, in horseradish sauce and mustard for instance.

- Make any allowance for slow acid equilibration in products such as sauce tartare.
- Apply to pasteurized non-emulsified sauces.

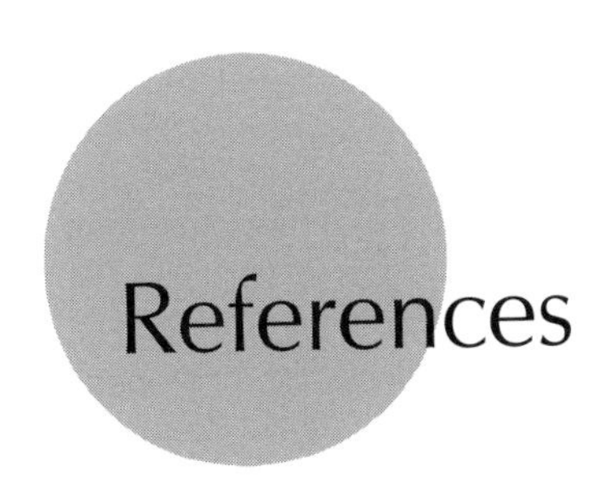

References

Aked, J. (2000) Fruits and vegetables. In: *The Stability and Shelf-life of Food* (eds D. Kilcast & P. Subramaniam), pp. 249–278. Woodhead Publishing, Cambridge, UK.

Andersen, H.J. & Rasmussen, M.A. (1992) Interactive packaging as protection against photodegradation of the colour of pasteurized, sliced ham. *International Journal of Food Science and Technology*, **27**(1), 1–8.

Anon. (1995) *The FRESHLINE Guide to Modified Atmosphere Packaging (MAP)*. Air Products plc, Basingstoke.

Anon. (1996) *The Food Safety Act 1990 and You* (PB2507). Advice from HM Government.

Baird-Parker, A.C. and Kilsby, D.C. (1987) Principles of predictive food microbiology. *Journal of Applied Bacteriology Symposium Supplement*, pp. 43S–49S.

Baranyi, J., Roberts, T.A. & McClure, P.J. (1993) A non-autonomous differential equation to model bacterial growth. *Food Microbiology*, **10**, 43–49.

Bell, L.N. (1997) Maillard reaction as influenced by buffer type and concentration. *Food Chemistry*, **59**(1), 143–147.

Bell, L.N. & Labuza, T.P. (1994) Aspartame stability in commercially sterilised flavoured dairy beverages. *Journal of Dairy Science*, **77**, 34–38.

Berry-Ottaway, P. (1993) Stability of vitamins in food. In: *Technology of Vitamins in Food* (ed P. Berry-Ottaway), pp. 90–113. Chapman & Hall, London.

Betts, G. & Everis, L. (2000) Shelf-life determination and challenge testing. In: *Chilled Foods – A Comprehensive Guide*, 2nd edn (eds M. Stringer & C. Dennis), pp. 259–285. Woodhead Publishing, Cambridge, UK.

Birollo, G.A., Reinheimer, J.A. & Vinderola, C.G. (2000) Viability of lactic acid microflora in different types of yoghurt. *Food Research International*, **33**, 799–805.

Blackburn, C. de W. (2000) Modelling shelf-life. In: *The Stability and Shelf-life of Food* (eds D. Kilcast & P. Subramaniam), pp. 55–78, Woodhead Publishing, Cambridge, UK.

Blanchfield, J.R. (1998) *Food and Drink Good Manufacturing Practice – A Guide to its Responsible Management*, 4th edn, Institute of Food Science and Technology (UK), London.

BSI (1999) *BS 7908: 1999 Packaging – Temperature and Time-temperature Indicator – Performance Specification and Reference Testing*. British Standards Institution, London.

Cairns, J.A. (1974) Measurement of permeability. In: *Packaging for Climatic Protection* (eds J.A. Cairns, C.R. Oswin & F.A. Paine), pp. 23–48, Newnes–Butterworths, London.

Cakebread, S.H. (1974) Confectionery ingredients: osmotic properties of carbohydrate solutions – XIII. *Confectionery Production*, **40**(3), 104, 106, 107, 109.

Carpenter, R.P., Lyon, D.H. & Hasdell, T.A. (2000) *Guidelines for Sensory Analysis in Food Product Development and Quality Control*, 2nd edn, Aspen Publishers, Gaithersburg, MD.

Cauvain, S. & Young, L. (2000) *Bakery Food Manufacture & Quality – Water Control and Effects*. Blackwell Science, Oxford.

CCFRA (1990) *Evaluation of Shelf Life for Chilled Foods*. Technical Manual No. 28. Campden and Chorleywood Food Research Association, Chipping Campden, Gloucestershire.

Charalambous, G. (ed) (1993) *Shelf Life Studies of Foods and Beverages – Chemical, Biological, Physical and Nutritional Aspects*, Elsevier Science, Amsterdam.

CIMSCEE (1991) *Code for the production of microbiologically safe and stable emulsified and non-emulsified sauces containing acetic acid*. Comité de Industries des Mayonnaises et Sauces Condimentaires de la Communauté Economique Européenne, Brussels.

Codex Commission on Food Hygiene (1997) *HACCP System and Guidelines for its Application*. Annex to CAC/RCP 1–1969, Rev. 3, in Codex Alimentarius Food Hygiene Basic Texts, Food and Agriculture Organisation of the United Nations, World Health Organization, Rome.

Conti, M.E. (1998) The content of heavy metals in food packaging paper boards: an atomic absorption spectroscopy investigation. *Food Research International*, **30**(5), 343–348.

Coupland, J.N. & McClements, D.J. (1996) Lipid oxidation in food emulsions. *Trends in Food Science and Technology*, **7**, 83–90.

Crawford, C. (1998) *The New QUID Regulations*. Chandos Publishing, Oxford.

Dalgaard, P. (2000) Fresh and lightly preserved seafood. In: *Shelf-Life Evaluation of Foods*, 2nd edn (eds C.M.D. Man & A.A. Jones), pp. 110–139. Aspen Publishers, Gaithersburg, MD.

Delamarre, S. and Batt, C.A. (1999) The microbiology and historical safety of margarine. *Food Microbiology*, **16**, 327–333.

Del Nobile, M.A., Mensitieri, G., Nicolais, L. & Masi, P. (1998) The influence of the thermal history on the shelf life of carbonated beverages bottled in plastic containers. *Journal of Food Engineering*, **34**, 1–13.

Dens, E.J. & Van Impe, J.F. (2001) On the need for another type of predictive model in structured foods. *International Journal of Food Microbiology*, **64**, 247–260.

Downham, A. and Collins, P. (2000) Colouring our foods in the last and next millennium. *International Journal of Food Science and Technology*, **35**, 5–22.

EC (1994) Commission Directive 94/54/EC amending Directive 79/112/EEC on the approximation of the laws of the Member States relating to the labelling, presentation and advertising of foodstuffs for sale to the ultimate consumers, OJ No. L300 of 23.11.1994, p. 14.

EC (1996) Commission Directive 96/21/EC amending Directive 94/54/EC concerning the compulsory indication on the labelling of certain foodstuffs of particulars other than those provided for in Council Directive 79/112/EEC, OJ No. L88 of 5.4.1996, p. 5.

EC (1999) Commission Directive 99/10/EC providing for derogation from the provisions of Article 7 of Council Directive 79/112/EEC as regards the labelling of foodstuffs, OJ No. L69 of 10.3.1999, p. 22.

EEC (1979) Council Directive 79/112/EEC on the approximation of the laws of the Member States relating to the labelling, presentation and advertising of foodstuffs for sale to the ultimate consumer (excluding provisions relating to net quantity, and except in relation to certain additives). OJ No. L33 of 8.2.1979.

EEC (1989) Commission Directive 89/109/EEC on the approximation of the laws of the Member States relating to materials and articles intended to come into contact with foodstuffs, OJ No L40 of 11.2.1989.

EEC (1992a) Commission Directive 92/1/EEC on the monitoring of temperatures in the means of transport, warehousing and storage of quick-frozen foodstuffs intended for human consumption, OJ No. L34 of 11.2.92, p. 28.

EEC (1992b) Commission Directive 92/2/EEC on the sampling procedure and the Community analysis for the official control of quick-frozen foods intended for human consumption, OJ No. L34 of 11.2.92, p. 30.

Ellis, M.J. and Man, C.M.D. (2000) The methodology of shelf-life determination. In: *Shelf-Life Evaluation of Foods*, 2nd edn (eds C.M.D. Man & A.A. Jones), pp. 23–33. Aspen Publishers, Gaithersburg, MD.

Evans, J.A. (1998) Consumer perceptions and practice in the handling of chilled foods. In: *Sous Vide and Cook–Chill Processing for the Food Industry* (ed S. Ghazala), pp. 312–360, Aspen Publishers, Gaithersburg, MD.

Evans, J.A., Stanton, J.I., Russell, S.L. & James, S.J. (1991) *Consumer Handling of Chilled Foods: A Survey of Time and Temperature Conditions.* MAFF Publications, London, p. 102.

Floros, J.D. & Gnanasekharan, V. (1993) Shelf life prediction of packaged food. In: *Shelf Life Studies of Food and Beverages – Chemical, Biological, Physical and Nutritional Aspects* (ed G. Charalambous), pp. 1081–1118, Elsevier Science, Amsterdam.

FSA (2000) *Food Law Guide*. Food Standards Agency (UK), London.

Fu, B. & Labuza, T.P. (1993) Shelf-life prediction: theory and application. *Food Control*, **4**(3), 125–133.

Gacula, M.C. (1975) The design of experiments for shelf-life study. *Journal of Food Science*, **40**, 399–403.

George, R.M. & Shaw, R. (1992) *A Food Industry Specification for Defining the Technical Standards and Procedures for the Evaluation of Temperature and Time–Temperature Indicators*. Technical Manual No. 35, Campden and Chorleywood Food Research Association, UK.

Gerba, C.P., Rose, J.B. & Haas, C.N. (1996) Sensitive populations: who is at the greatest risk? *International Journal of Food Microbiology*, **30**, 113–123.

Gilbert, R.J., de Louvois, J., Donovan, T., Little, C., Nye, K., Bibeiro, C.D., Richards, J., Roberts, D. & Bolton, F.J. (2000) Guidelines for the microbiological quality of some ready-to-eat foods sampled at the point of sale. *Communicable Disease and Public Health*, **3**(3), 163–167.

Gill, C.O. (1996) Cold storage temperature fluctuations and predicting microbial growth. *Journal of Food Protection*, **59**, 43–47.

Goddard, M.R. (2000) The storage of thermally processed foods in containers other than cans. In: *Shelf-Life Evaluation of Foods*, 2nd edn (eds C.M.D. Man & A.A. Jones), pp. 197–210. Aspen Publishers, Gaithersburg, MD.

Goodburn, K. (2000) Legislation. In: *Chilled Foods – A Comprehensive Guide*, 2nd edn (eds M. Stringer & C. Dennis), pp. 451–473. Woodhead Publishing, Cambridge, UK.

Gould, G.W. (1996) Methods of preservation and extension of shelf life. *International Journal of Food Microbiology*, **33**, 51–64.

Gould, G.W. (1999) Sous vide foods: conclusions of an ECFF botulism working party. *Food Control*, **10**, 47–51.

Hamilton, R.J. (1994) The chemistry of rancidity in foods. In: *Rancidity in Foods*, 3rd edn (eds J.C. Allen and R.J. Hamilton), pp. 1–21, Blackie Academic & Professional, London.

Hansen, L.T., Røntved, S.D. & Huss, H.H. (1998) Microbiological quality and shelf life of cold-smoked salmon from three different processing plants. *Food Microbiology*, **15**, 137–150.

HMSO (1970) *Measurement of Humidity*. Notes on Applied Science No. 4, Her Majesty's Stationery Office, London.

HMSO (1978) *The Coffee and Coffee Products Regulations* (SI 1978/1420). Her Majesty's Stationery Office, London.

HMSO (1990) *Food Safety Act*. Her Majesty's Stationery Office, London.

HMSO (1992) *The Tin in Food Regulations* (SI 1992/496). Her Majesty's Stationery Office, London.

HMSO (1994) *The Quick-frozen Foodstuffs (Amendment) Regulations* (SI 1994/298). Her Majesty's Stationery Office, London.

HMSO (1995a) *The Dairy Products (Hygiene) Regulations* (SI 1995/1086). Her Majesty's Stationery Office, London.

HMSO (1995b) *Food Safety (Temperature Control) Regulations 1995* (SI 1995/2200). Her Majesty's Stationery Office, London.

HMSO (1996) *The Food Labelling Regulations* (SI 1996/1499). Her Majesty's Stationery Office, London.

Holley, R.A. (1997) Impact of slicing hygiene upon shelf life and distribution of spoilage bacteria in vacuum packaged cured meats. *Food Microbiology*, **14**, 201–211.

Holzapfel, W.H., Haberer, P., Snel, J., Schillinger, U. & Huis in't Veld, J.H.J. (1998) Overview of gut flora and probiotics. *International Journal of Food Microbiology*, **41**, 85–101.

Howarth, J.A.K. (2000) Ready-to-eat breakfast cereals. In: *Shelf-Life Evaluation of Foods*, 2nd edn (eds C.M.D. Man & A.A. Jones), pp. 182–196. Aspen Publishers, Gaithersburg, MD.

Huis in't Veld, J.H.J. (1996) Microbial and biochemical spoilage of foods: an overview. *International Journal of Food Microbiology*, **33**, 1–18.

ICMSF (1988) *Microorganisms in Foods 4 – Application of the Hazard Analysis Critical Control Point (HACCP) System to Ensure Microbiological Safety and Quality*. International Commission on Microbiological Specifications for Foods, Blackwell Scientific, Oxford.

ICMSF (1996) *Microorganisms in Foods 5 – Microbiological Specifications of Food Pathogens*. International Commission on Microbiological Specifications for Foods, Blackie Academic & Professional, London.

ICMSF (1998) *Microorganisms in Foods 6 – Microbial Ecology of Food Commodities*. International Commission on Microbiological Specifications for Foods, Blackie Academic & Professional, London.

IFST (1993) *Shelf Life of Foods – Guidelines for its Determination and Prediction*. Institute of Food Science & Technology (UK), London.

IFST (1997a) Cryptosporidium. IFST Position Statement, *Food Science and Technology Today*, **11**(1), 46–48.

IFST (1997b) Foodborne viral infections. IFST Position Statement, *Food Science and Technology Today*, **11**(1), 49–51.

IFST (1998) *Microbiological Food Safety for Children and Vulnerable Groups*. IFST Position Statement, October. Institute of Food Science and Technology (UK), London.

IFST (1999) *Development and Use of Microbiological Criteria for Foods*. Institute of Food Science and Technology (UK), London.

IFT (1981) Sensory evaluation guide for testing food and beverage products. *Food Technology*, **35**, 50–57.

Johnston, D.E. (1994) High pressure – a new dimension to food processing. *Chemistry and Industry*, No. 14, July, 499–501.

Jones, A.A. (2000a) Ambient-stable sauces and pickles. In: *Shelf-Life Evaluation of Foods*, 2nd edn (eds C.M.D. Man & A.A. Jones), pp. 140–156. Aspen Publishers, Gaithersburg, MD.

Jones, H.P. (2000b) Ambient packaged cakes. In: *Shelf-Life Evaluation of Foods*, 2nd edn (eds C.M.D. Man & A.A. Jones), pp. 211–226. Aspen Publishers, Gaithersburg, MD.

Jung, M.Y., Kim, S.K. & Kim, S.Y. (1995) Riboflavin-sensitised photooxidation of ascorbic acid: kinetics and amino acid effects. *Food Chemistry*, **53**, 397–403.

Kanner, J., Ben-Gera, I. & Berman, S. (1980) Nitric-oxide myoglobin as an inhibitor of lipid oxidation. *Lipids*, **15**, 944–948.

Kanner, J., Harel, S., Shagalovich, J. & Berman, S. (1984) Antioxidative effect of nitrite in cured meat products: nitric oxide–iron complexes of low molecular weight. *Journal of Agricultural and Food Chemistry*, **32**, 512–515.

Katsaras, K. & Leistner, L. (1991) Distribution and development of bacterial colonies in fermented sausages. *Biofouling*, **5**, 115–124.

Kilcast, D. (2000) Sensory evaluation methods for shelf-life assessment. In: *The Stability and Shelf-life of Foods* (eds D. Kilcast and P. Subramaniam), pp. 79–105, Woodhead Publishing, Cambridge, UK.

Kilcast, D. & Subramaniam, P. (eds) (2000) *The Stability and Shelf-life of Food*, Woodhead Publishing, Cambridge, UK.

Kim, S.K., Jung, M.Y. & Kim, S.Y. (1997) Photodecomposition of aspartame in aqueous solutions. *Food Chemistry*, **59**(2), 273–278.

Knorr, D. (1998) Technology aspects related to microorganisms in functional foods. *Trends in Food Science and Technology*, **9**, 295–306.

Labuza, T.P. (1982) *Shelf-life Dating of Foods*. Food and Nutrition Press, Westport, CT.

Labuza, T.P. & Hyman, C.R. (1998) Moisture migration and control in multi-domain foods. *Trends in Food Science and Technology*, **9**, 47–55.

Labuza, T.P. & Riboh, D. (1982) Theory and application of Arrhenius kinetics to the prediction and nutrient losses in foods. *Food Technology*, **10**, 66, 68, 70, 72, 74.

Labuza, T.P. and Schmidl, M.K. (1985) Accelerated shelf-life testing of foods. *Food Technology*, **9**, 57–62, 64, 134.

Labuza, T.P. & Schmidl, M.K. (1988) Use of sensory data in the shelf life testing of foods: principles and graphical methods for evaluation. *Cereal Foods World*, **33**(2), 193–204.

Lawless, H.T. & Heymann, H. (1998) *Sensory Evaluation of Food, Principles and Practices*. Chapman & Hall, New York.

Leistner, L. (1992) Food preservation by combined methods. *Food Research International*, **25**, 151–158.

Leistner, L. (2000) Minimally processed, ready-to-eat, and ambient-stable meat. In: *Shelf-life Evaluation of Foods* (eds C.M.D. Man & A.A. Jones), pp. 242–262, Aspen Publishers, Gaithersburg, MD.

McDonald, K. & Sun, D.W. (1999) Predictive food microbiology for the meat industry: a review. *International Journal of Food Microbiology*, **52**, 1–27.

McMeekin, T.A., Olley, J.N., Ross, T. & Ratkowsky, D.A. (1993) *Predictive Microbiology: Theory and Application*. Research Studies Press, Somerset.

McMeekin, T.A. & Ross, T. (1996) Shelf-life prediction: status and future possibilities. *International Journal of Food Microbiology*, **33**, 65–83.

McMurrough, I., Madigan, D., Kelly, R. & O'Rourke, T. (1999) Haze formation and shelf-life prediction for lager beer. *Food Technology*, **53**(1), 58–62.

Man, C.M.D. (2000) Potato chips and savory snacks. In: *Shelf-life Evaluation of Foods* (eds C.M.D. Man & A.A. Jones), pp. 157–168, Aspen Publishers, Gaithersburg, MD.

Man, C.M.D. & Jones, A.A. (eds) (2000) *Shelf-Life Evaluation of Foods*, 2nd edn, Aspen Publishers, Gaithersburg, MD.

Martinez, M.V. & Whitaker, J.R. (1995) The biochemistry and control of enzymatic browning. *Trends in Food Science and Technology*, **6**, 195–200.

Matthews, A.C. (1995) Managing stability in the beverage industry. *Food Technology International Europe*, 77, 78, 80–81.

Mermelstein, N.H. (1998) High pressure processing begins, *Food Technology*, **52**(6), 104–106.

Mizrahi, S. (2000) Accelerated shelf-life tests. In: *The Stability and Shelf-life of Food* (ed. D. Kilcast & P. Subramaniam), pp. 107–128. Woodhead Publishing, Cambridge, UK.

Mossel, D.A.A. (1971) Physiological and metabolic attributes of microbial groups associated with foods. *Journal of Applied Bacteriology*, **34**, 95–118.

Mossel, D.A.A., Corry, J.E.L., Struijk, C.B. & Baird, R.M. (1995) *Essentials of the Microbiology of Foods: a Textbook for Advanced Studies*. pp. 175–214. Wiley, UK.

Mröhs, A. (2000) Durability indication: European Union. In: *Food Labelling* (ed J. Ralph Blanchfield), pp. 101–109. Woodhead Publishing, Cambridge, UK.

Nielsen, T. & Jägerstad, M. (1994) Flavour scalping by food packaging. *Trends in Food Science and Technology*, **5**, 353–356.

Notermans, S., in't Veld, P., Wijtzes, T. & Mead, G.C. (1993) A user's guide to microbial challenge testing for ensuring the safety and stability of food products. *Food Microbiology*, **10**, 145–157.

O'Mahony, M. (1986) *Sensory Evaluation of Food. Statistical Methods and Procedures*. Marcel Dekker, New York.

Pacquette, C.L. (1998) *Stability of Selected Water-Soluble Vitamins in Model Systems*, PhD Thesis, South Bank University, London.

Painter, T.J. (1998) Carbohydrate polymers in food preservation: an integrated view of the Maillard reaction with special reference to discoveries of preserved foods in sphagnum-dominated peat bogs. *Carbohydrate Polymer*, **36**(4), 335–347.

Panisello, P.J. & Quantick, P.C. (1998) Hazard analysis critical control point and its implementation: the need for an international microbiological hazard database. *Food Science and Technology Today*, **12**(3), 130–133.

Petersen, M.A., Tønder, D. & Poll, L. (1998) Comparison of normal and accelerated storage of commercial orange juice – changes in flavour and content of volatile compounds. *Food Quality and Preference*, **9**, 43–51.

Ratkowsky, D.A., Olley, J., McMeekin, T.A. & Ball, A. (1982) Relationship between temperature and growth rate of bacterial cultures. *Journal of Bacteriology*, **149**, 1–5.

Roberts, T.A. & Gibson, A.M. (1986) Interactions of food components affecting microbial growth. In: *Interactions of Food Components* (ed G.G. Birch & M.G. Lindley), pp. 131–142. Elsevier Applied Science, Essex.

Robins, M., Brocklehurst, T. & Wilson, P. (1994) Food structure and the growth of pathogenic bacteria. *Food Technology International Europe*, 31, 32, 34–36.

Roig, M.G., Bello, J.F., Rivera, Z.S. & Kennedy, J.F. (1999) Studies on the occurrence of non-enzymatic browning during storage of citrus juice. *Food Research International*, **32**, 609–619.

Rose, S.A. (1987) *Guidelines for Microbiological Challenge Testing.* Technical Manual No. 20. Campden Food & Drink Research Association, Chipping Campden, Gloucestershire.

Sanders, T.A.B. (1994) Nutritional aspects of rancidity. In: *Rancidity in Foods*, 3rd edn, (eds J.C. Allen & R.J. Hamilton), pp. 128–140, Blackie Academic & Professional, London.

Shortt, C. (1999) The probiotic century: historical and current perspective. *Trends in Food Science and Technology*, **10**, 411–417.

Sinell, H.-J. (1995) Control of food-borne infections and intoxications. *International Journal of Food Microbiology*, **25**, 209–217.

Singh, R.P. (2000) Scientific principles of shelf-life evaluation. In: *Shelf-life Evaluation of Foods* (eds C.M.D. Man and A.A. Jones), pp. 3–22. Aspen Publishers, Gaithersburg, MD.

Sizer, C. (2000) Engineering a safer food supply. *Chemistry and Industry*, No. 19, October, 637–640.

Skog, K.I., Johansson, M.A.E. & Jagerstad, M.I. (1998) Carcinogenic heterocyclic amines in model systems and cooked foods: a review on formation, occurrence and intake. *Food Chemical Toxicology*, **36**, 879–896.

Stone, H. & Sidel, J.L. (1993) *Sensory Evaluation Practices*. Academic Press, FL.

Taoukis, P.S., Labuza, T.P. & Saguy, I.S. (1997) Kinetics of food deterioration and shelf-life prediction. In: *Handbook of Food Engineering Practice* (eds K.J. Valentas, E. Rotstein & R.P. Singh), pp. 361–403. CRC Press, Boca Raton, FL.

Turner, T.A. (1998) *Canmaking – The Technology of Metal Protection and Decoration.* Blackie Academic & Professional, London.

Tuynenburg Muys, G. (1965) Microbiological quality of edible emulsions during manufacture and storage. *Chemistry and Industry,* No. 13, July, 1245–1250.

Tuynenburg Muys, G. (1969) Microbiology of margarine. *Process Biochemistry,* **4**, 31–34.

Tuynenburg Muys, G. (1971) Microbial safety in emulsions. *Process Biochemistry,* **6**, 25–28.

Untermann, F. (1998) Microbial hazards in food. *Food Control,* **9**(2–3), 119–126.

Vermeiren, L., Devlieghere, F., van Beest, M., de Kruijf, N. & Debevere, J. (1999) Developments in the active packaging of foods. *Trends in Food Science and Technology,* **10**, 77–86.

Walker, J.R.L. & Ferrar, P.H. (1995) The control of enzymic browning in foods. *Chemistry and Industry,* No. 20, October, 836–839.

Walker, S.J. (2000) The principles and practice of shelf-life prediction for microorganisms. In: *Shelf-life Evaluation of Foods,* 2nd edn (eds C.M.D. Man & A.A. Jones), pp. 34–41. Aspen Publishers, Gaithersburg, MD.

Whiting, R.C. (1995) Microbial modelling in foods. *Critical Reviews in Food Science and Nutrition,* **35**, 467–494.

Whiting, R.C. & Buchanan, R.L. (1993) A classification of models for predictive microbiology. *Food Microbiology,* **10**, 175–177.

Wilson, P.D.G. & Hibberd, D.J. (2000) The prediction of pH in complex foods. *Food Science and Technology Today,* **14**(2), 72–75.

Index